AUTOMATED CONTINUOUS PROCESS CONTROL

AUTOMATED CONTINUOUS PROCESS CONTROL

CARLOS A. SMITH
Chemical Engineering Department
University of South Florida

A Wiley-Interscience Publication
JOHN WILEY & SONS, INC.

For ordering and customer service information please call 1-800-CALL-WILEY.

Library of Congress Cataloging-in-Publication Data Is Available

ISBN 0-471-21578-3

Printed in the United States of America.

10 9 8 7 6 5 4 3 2 1

This work is dedicated to the Lord our God, for his daily blessings make all our work possible.

To the old generation: Mami, Tim, and Cristina Livingston, and Carlos and Jennifer Smith.

To the new generation: Sophia Cristina Livingston and Steven Christopher Livingston.

To my dearest homeland, Cuba.

CONTENTS

PREFACE   **xi**

1 INTRODUCTION   **1**

   1-1 Process Control System / 1
   1-2 Important Terms and Objective of Automatic Process Control / 3
   1-3 Regulatory and Servo Control / 4
   1-4 Transmission Signals, Control Systems, and Other Terms / 5
   1-5 Control Strategies / 6
       1-5.1 Feedback Control / 6
       1-5.2 Feedforward Control / 8
   1-6 Summary / 9

2 PROCESS CHARACTERISTICS   **11**

   2-1 Process and Importance of Process Characteristics / 11
   2-2 Types of Processes / 13
   2-3 Self-Regulating Processes / 14
       2-3.1 Single-Capacitance Processes / 14
       2-3.2 Multicapacitance Processes / 24
   2-4 Transmitters and Other Accessories / 28
   2-5 Obtaining Process Characteristics from Process Data / 29
   2-6 Questions When Performing Process Testing / 32
   2-7 Summary / 33
      Reference / 33
      Problems / 34

3 **FEEDBACK CONTROLLERS**   38

   3-1 Action of Controllers / 38

   3-2 Types of Feedback Controllers / 40

      3-2.1 Proportional Controller / 40

      3-2.2 Proportional–Integral Controller / 44

      3-2.3 Proportional–Integral–Derivative Controller / 48

      3-2.4 Proportional–Derivative Controller / 50

   3-3 Reset Windup / 50

   3-4 Tuning Feedback Controllers / 53

      3-4.1 Online Tuning: Ziegler–Nichols Technique / 53

      3-4.2 Offline Tuning / 54

   3-5 Summary / 60

      References / 60

      Problems / 60

4 **CASCADE CONTROL**   61

   4-1 Process Example / 61

   4-2 Implementation and Tuning of Controllers / 65

      4-2.1 Two-Level Cascade Systems / 65

      4-2.2 Three-Level Cascade Systems / 68

   4-3 Other Process Examples / 69

   4-4 Closing Comments / 72

   4-5 Summary / 73

      References / 73

5 **RATIO, OVERRIDE, AND SELECTIVE CONTROL**   74

   5-1 Signals and Computing Algorithms / 74

      5-1.1 Signals / 74

      5-1.2 Programming / 75

      5-1.3 Scaling Computing Algorithms / 76

      5-1.4 Significance of Signals / 79

   5-2 Ratio Control / 80

   5-3 Override, or Constraint, Control / 88

   5-4 Selective Control / 92

   5-5 Designing Control Systems / 95

   5-6 Summary / 110

      References / 111

      Problems / 112

6 **BLOCK DIAGRAMS AND STABILITY**   127

   6-1 Block Diagrams / 127

   6-2 Control Loop Stability / 132

6-2.1 Effect of Gains / 137

6-2.2 Effect of Time Constants / 138

6-2.3 Effect of Dead Time / 138

6-2.4 Effect of Integral Action in the Controller / 139

6-2.5 Effect of Derivative Action in the Controller / 140

6-3 Summary / 141

Reference / 141

7 FEEDFORWARD CONTROL   **142**

7-1 Feedforward Concept / 142

7-2 Block Diagram Design of Linear Feedforward Controllers / 145

7-3 Lead/Lag Term / 155

7-4 Extension of Linear Feedforward Controller Design / 156

7-5 Design of Nonlinear Feedforward Controllers from Basic Process Principles / 161

7-6 Closing Comments on Feedforward Controller Design / 165

7-7 Additional Design Examples / 167

7-8 Summary / 172

References / 173

Problem / 173

8 DEAD-TIME COMPENSATION   **174**

8-1 Smith Predictor Dead-Time Compensation Technique / 174

8-2 Dahlin Controller / 176

8-3 Summary / 179

References / 179

9 MULTIVARIABLE PROCESS CONTROL   **180**

9-1 Pairing Controlled and Manipulated Variables / 181

9-1.1 Obtaining Process Gains and Relative Gains / 186

9-1.2 Positive and Negative Interactions / 189

9-2 Interaction and Stability / 191

9-3 Tuning Feedback Controllers for Interacting Systems / 192

9-4 Decoupling / 194

9-4.1 Decoupler Design from Block Diagrams / 194

9-4.2 Decoupler Design from Basic Principles / 196

9-5 Summary / 197

References / 197

Problem / 198

Appendix A CASE STUDIES   **199**

Case 1: Ammonium Nitrate Prilling Plant Control System / 199

Case 2: Natural Gas Dehydration Control System / 200

Case 3: Sodium Hypochlorite Bleach Preparation Control System / 201

Case 4: Control System in the Sugar Refining Process / 202

Case 5: Sulfuric Acid Process / 204

Case 6: Fatty Acid Process / 205
      Reference / 207

Appendix B   PROCESSES FOR DESIGN PRACTICE   **208**

Installing the Programs / 208

Process 1: NH_3 Scrubber / 208

Process 2: Catalyst Regenerator / 211

Process 3: Mixing Process / 213

INDEX   **215**

PREFACE

This book was written over a number of years while teaching short courses to industry. Most of the participants were graduate engineers, and a few were instrument technicians. For the engineers, the challenge was to show them that the control theory most of them heard in college is indeed the basis for the practice of process control. For the technicians, the challenge was to teach them the practice of process control with minimum mathematics. The book does not emphasize mathematics, and a serious effort has been made to explain, using readable language, the meaning and significance of every term used: that is, what the term is telling us about the process, about the controller, about the control performance, and so on.

The book assumes that the reader does not know much about process control. Accordingly, Chapter 1 presents the very basics of process control. While several things are presented in Chapter 1, the main goals of the chapter are (1) to present why process control is needed, (2) to present the basic components of a control system, (3) to define some terms, and (4) to present the concept of feedback control with its advantages, disadvantages, and limitations.

To do good process control there are at least three things the practitioner should know and fully understand: (1) the process, (2) the process, and (3) the process! Chapter 2 presents a discussion of processes from a very physical point of view. Everything presented in this chapter is used extensively in all remaining chapters.

Chapter 3 presents a discussion of feedback controllers, and specifically, the workhorse in the process industry: the PID controller. A significant effort is made to explain each tuning parameter in detail as well as the different types of controllers, with their advantages and disadvantages. In the chapter we describe how to tune, adjust, or adapt the controller to the process. Finally, we discuss the important topics of reset windup, tracking, and tuning flow and level loops. Throughout the presentation, the use of distributed control systems (DCSs) is stressed. Problems are presented at the end of Chapters 2 and 3 to practice what was presented.

As discussed in Chapter 1, feedback control has the limitation that in some cases it does not provide enough control performance. In these cases some other control strategy is needed to obtain the control performance required. What is usually done is to provide assistance to feedback control; feedback control is never removed. Cascade control is a common strategy to improve simple feedback control. In Chapter 4 we present the concept and implementation of cascade control.

In Chapter 5 we describe ratio, override (or constraint), and selective control. To implement these strategies, some computing power is needed. The chapter starts with a presentation of how DCSs handle signals as they enter the system and a description of different programming techniques and computing power. Ratio, override, and selective control are presented using examples. The chapter ends with some hints on how to go about designing these strategies. Many problems are given at the end of the chapter.

Once feedback and cascade control have been presented, it is worthwhile to discuss the important subject of control system stability. Chapter 6 starts with the subject of block diagram and continues with the subject of stability. Block diagrams are used in subsequent chapters to explain the implementation of other control strategies. Stability is presented from a very practical point of view without dealing much with mathematics. It is important for the practitioner to understand how each term in the control system affects the stability of the system.

The detrimental effect of dead time on the stability of a control system is presented in Chapter 6. Chapter 7 is devoted exclusively to feedforward control. Various ways to design and implement this important compensation strategy and several examples are presented. Several techniques to control processes with long dead times are described in Chapter 8, and multivariable process control in Chapter 9. Appendix A provides some process examples to design the control strategies for an entire process. Finally, Appendix B describes the processes presented in the compact disk (CD). These processes have been used for many years to practice tuning feedback and cascade controllers as well as designing feedforward controllers.

The author believes that to practice industrial process control (as opposed to "academic" process control), there is generally no need for advanced mathematics. The author is also aware that the reader is interested in learning "just enough theory" to practice process control. The main concern during the writing of this manuscript has been to present the reader with the benefits obtained with good control, and in doing so, to motivate him or her to learn more about the subject. We hope you do so, and now wish you good controlling!

It is impossible to write a book like this one without receiving help and encouragement from other people. The author would first like to acknowledge the encouragement received from the hundreds of engineers and technicians who have attended the short courses and offered suggestions and examples. The author would also like to sincerely thank his friends, colleagues, and most outstanding chemical engineers, J. Carlos Busot and Armando B. Corripio (coauthor of *Principles and Practice of Automatic Process Control*). Their friendship, human quality, professional quality, and ability to frustrate the author have had a great positive impact in my life. Thanks to both of you! ABC also provided the material presented in Section 8-2. The author also remembers very dearly his former student, the late Dr. Daniel Palomares, for his contributions to the simulations presented in the CD accompanying this book. Finally, the author would like to thank his graduate student and friend, Dr. Marco Sanjuan. Marco's friendship, support, and continuous encouragement have made these past years a tremendous pleasure. Marco also put the final touches to the CD.

Tampa, FL   CARLOS A. SMITH, PH.D., P.E.
2001

CHAPTER 1

INTRODUCTION

Automatic process control is concerned with maintaining process variables, temperatures, pressures, flows, compositions, and the like, at a desired operating value. As we shall see in the ensuing pages, processes are dynamic in nature. Changes are always occurring, and if actions are not taken, the important process variables—those related to safety, product quality, and production rates—will not achieve design conditions.

1-1 PROCESS CONTROL SYSTEM

To fix ideas, let us consider a heat exchanger in which a process fluid is heated by condensing steam; the process is sketched in Fig. 1-1.1. The purpose of this unit is to heat the process fluid from some inlet temperature, $T_i(t)$, up to a desired outlet temperature, $T(t)$. The energy gained by the process fluid is provided by the latent heat of condensation of the steam.

In this process many variables can change, causing the outlet temperature to deviate from its desired value. If this happens, some action must be taken to correct for this deviation. The objective is to maintain the outlet process temperature at its desired value. One way to accomplish this objective is to first measure the temperature, $T(t)$, compare it to its desired value, and based on this comparison, decide what to do to correct for any deviation. The steam valve can be manipulated to correct for the deviation. That is, if the temperature is above its desired value, the steam valve can be throttled back to cut the steam flow (energy) to the heat exchanger. If the temperature is below its desired value, the steam valve could be opened more to increase the steam flow to the exchanger. The operator can do all of this manually, and since the procedure is fairly straightforward, it should present no problem. However, there are several problems with this *manual process control*. First, the job requires that the operator look frequently at the temperature to take

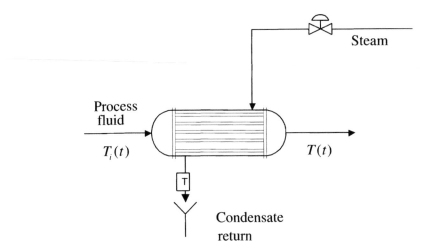

Figure 1-1.1   Heat exchanger.

corrective action whenever it deviates from the value desired. Second, different operators would make different decisions as to how to move the steam valve, resulting in inconsistent operation. Third, since in most process plants hundreds of variables must be maintained at a desired value, this correction procedure would require a large number of operators. Consequently, we would like to accomplish this control automatically. That is, we would like to have systems that control the variables without requiring intervention from the operator. This is what is meant by *automatic process control*.

To accomplish this objective, a control system must be designed and implemented. A possible control system and its basic components are shown in Fig. 1-1.2. The first thing to do is to measure the outlet temperature of the process stream. This is done by a sensor (thermocouple, resistance temperature device, filled system thermometers, thermistors, etc.). Usually, this sensor is connected physically to a transmitter, which takes the output from the sensor and converts it to a signal strong enough to be transmitted to a controller. The controller then receives the signal, which is related to the temperature, and compares it with the value desired. Depending on this comparison, the controller decides what to do to maintain the temperature at its desired value. Based on this decision, the controller sends a signal to the final control element, which in turn manipulates the steam flow. This type of control strategy is known as *feedback control*.

The preceding paragraph presented the three basic components of all control systems:

1. *Sensor/transmitter*: also often called the *primary* and *secondary elements*
2. *Controller*: the "brain" of the control system
3. *Final control element*: often a control valve, but not always.

Other common final control elements are variable-speed pumps, conveyors, and electric motors.

The importance of these components is that they perform the three basic operations that must by present in every control system:

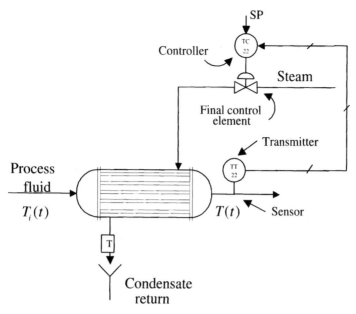

Figure 1-1.2   Heat exchanger control loop.

1. *Measurement* (*M*). Measuring the variable to be controlled is usually done by the combination of sensor and transmitter.
2. *Decision* (*D*). Based on the measurement, the controller decides what to do to maintain the variable at its desired value.
3. *Action* (*A*). As a result of the controller's decision, the system must then take an action. This is usually accomplished by the final control element.

These three operations, M, D, A, are always present in every type of control system. It is imperative, however, that the three operations be in a loop. That is, based on the measurement, a decision is made, and based on this decision, an action is taken. *The action taken must come back and affect the measurement*; otherwise, there is a major flaw in the design and control will not be achieved; when the action taken does not affect the measurement, an open-loop condition exists. The decision making in some systems is rather simple, whereas in others it is more complex; we look at many of them in this book.

1-2   IMPORTANT TERMS AND OBJECTIVE OF AUTOMATIC PROCESS CONTROL

At this time it is necessary to define some terms used in the field of automatic process control. The first term is *controlled variable*, which is the variable that must be maintained, or controlled, at some desired value. In the preceding discussion, the process outlet temperature, $T(t)$, is the controlled variable. Sometimes the terms *process variable* and/or *measurement* are also used to refer to the controlled variable. The *set point* is the desired value of the controlled variable. Thus the job of a control system is to maintain the controlled variable at its set point. The *manipulated variable* is the variable used to maintain the controlled variable at its set point. In the example, the steam valve position is the manipulated variable. Finally, any variable that causes the controlled variable to deviate away from the set point is defined as a *disturbance* or *upset*. In most processes there are a number of different disturbances. As an example, in the heat exchanger shown in Fig. 1-1.2, possible disturbances are the inlet process temperature $T_i(t)$, the process flow $f(t)$, the energy content of the steam, ambient conditions, process fluid composition, and fouling. It is important to understand that disturbances are always occurring in processes. Steady state is not the rule; transient conditions are very common. It is because of these disturbances that automatic process control is needed. If there were no disturbances, design operating conditions would prevail and there would be no necessity of continuously "monitoring" the process.

With these terms defined, we can simply state the following: *The objective of an automatic process control system is to adjust the manipulated variable to maintain the controlled variable at its set point in spite of disturbances.*

It is wise to enumerate some of the reasons why control is important. These are based on our industrial experience and we would like to pass them on to the reader. They may not be the only ones, but we feel they are the most important.

1. Prevent injury to plant personnel, protect the environment by preventing emissions and minimizing waste, and prevent damage to the process equipment. *Safety* must always be in everyone's mind; it is the single most important consideration.
2. Maintain product quality (composition, purity, color, etc.) on a continuous basis and with minimum cost.
3. Maintain plant production rate at minimum cost.

So it can be said that the reasons for automation of process plants are to provide safety and at the same time maintain desired product quality, high plant throughput, and reduced demand on human labor.

The following additional terms are also important. *Manual control* refers to the condition in which the controller is disconnected from the process. That is, the controller is not making the decision as to how to maintain the controlled variable at the set point. It is up to the operator to manipulate the signal to the final control element to maintain the controlled variable at the set point. *Automatic* or *closed-loop control* refers to the condition in which the controller is connected to the process, comparing the set point to the controlled variable, and determining and taking corrective action.

1-3 REGULATORY AND SERVO CONTROL

In some processes the controlled variable deviates from the set point because of disturbances. *Regulatory control* refers to systems designed to compensate for these disturbances. In some other instances the most important disturbance is the set point itself. That is, the set point may be changed as a function of time (typical of this is a batch reactor where the temperature must follow a desired profile), and therefore the controlled variable must follow the set point. *Servo control* refers to control systems designed for this purpose.

Regulatory control is far more common than servo control in the process industries. However, the basic approach to designing them is essentially the same. Thus the principles discussed in this book apply to both cases.

1-4 TRANSMISSION SIGNALS, CONTROL SYSTEMS, AND OTHER TERMS

There are three principal types of signals in use in the process industries. The pneumatic signal, or air pressure, ranges normally between 3 and 15 psig. The usual representation in piping and instrument diagrams (P&IDs) for pneumatic signals is ——//——//——. The electrical signal ranges normally between 4 and 20 mA; 1 to 5 V or 0 to 10 V are also used. The usual representation for this signal is a series of dashed lines such as – — — —. The third type of signal is the digital, or discrete, signal (zeros and ones); a common representation is ○——○——○——. In these notes we show signals as ——/——/—— (as shown in Fig. 1-1.2), which is the representation proposed by the Instrument Society of America (ISA) when a control concept is shown without concern for specific hardware. Generally, we refer to signals as a percent, 0 to 100%, as opposed to psig or mA. That is, 0 to 100% is equivalent to 3 to 15 psig or 4 to 20 mA.

It will help in understanding control systems to realize that signals are used by devices (transmitters, controllers, final control elements, etc.) to communicate. That is, signals are used to convey information. The signal from the transmitter to the controller is used by the transmitter to inform the controller of the value of the controlled variable. It is not the measurement in engineering units, but rather, a mA, psig, volt, or other signal that is proportional to the measurement. The relationship to the measurement depends on the calibration of the sensor/transmitter. The controller uses its output signal to indicate to the final control element what to do (i.e., how much to open if it is a valve, how fast to run if it is a variable-speed pump, etc.). Thus every signal is related to some physical quantity that makes sense from an engineering point of view. The signal from the temperature transmitter in Fig. 1-1.2 is related to the outlet temperature, and the signal from the controller is related to the steam valve position.

It is often necessary to change one type of signal into another type. A *transducer* or *converter* does this. For example, there may be a need to change from an electrical signal, mA, to a pneumatic signal, psig. This is done by the use of a current (I) to pneumatic (P) transducer (I/P). The input signal may be 4 to 20 mA and the output 3 to 15 psig. An analog-to-digital (A to D) converter changes from an mA or volt signal to a digital signal. There are many other types of transducers: digital to analog (D to A), pneumatic to current (P/I), voltage to pneumatic (E/P), pneumatic to voltage (P/E), and so on.

The term *analog* refers to the controller, or any other instrument, which is pneumatic, electrical, hydraulic, or mechanical. Most controllers however, are *computer-based*, or *digital*. By computer-based we don't necessarily mean a mainframe computer but rather, anything starting from a microprocessor. In fact, most controllers are microprocessor-based.

1-5 CONTROL STRATEGIES

1-5.1 Feedback Control

The control scheme shown in Fig. 1-1.2 is referred to as *feedback control*, also called a *feedback control loop*. One must understand the working principles of feedback control to recognize its advantages and disadvantages; the heat exchanger control loop shown in Fig. 1-1.2 is presented to foster this understanding.

If the inlet process temperature decreases, thus creating a disturbance, its effect must propagate through the heat exchanger before the outlet temperature decreases. Once this temperature changes, the signal from the transmitter to the controller also changes. It is then that the controller becomes aware that a deviation from set point has occurred and that it must compensate for the disturbance by manipulating the steam valve. The controller then signals the valve to increase its opening and thus increase the steam flow. Figure 1-5.1 shows graphically the effect of the disturbance and the action of the controller.

It is instructive to note that at first the outlet temperature decreases, because of the decrease in inlet temperature, but it then increases, even above the set point and continues to oscillate until it finally stabilizes. This oscillatory response is typical of feedback control and shows that it is essentially a trial and error operation. That is, when the controller notices that the outlet temperature has decreased below the set point, it signals the valve to open, but the opening is more than required. Therefore, the outlet temperature increases above the set point. Noticing this, the controller signals the valve to close again somewhat to bring the temperature back down. This trial and error continued until the temperature reached and stayed at set point.

The *advantage* of feedback control is that it is a very simple technique that compensates for all disturbances. Any disturbance affects the controlled variable, and once this variable deviates from the set point, the controller changes its output to return the controlled variable to set point. The feedback control loop does not know, nor does it care, which disturbance enters the process. It only tries to maintain the controlled variable at set point and in so doing compensates for all disturbances. The feedback controller works with minimum knowledge of the process. In fact, the only information it needs is in which direction to move. How much to move is usually adjusted by trial and error. The *disadvantage* of feedback control is that it can compensate for a disturbance only after the controlled variable has deviated from the set point. That is, the disturbance must propagate through the entire process before the feedback control scheme can compensate for it.

The job of the engineer is to design a control scheme that will maintain the controlled variable at its set point. Once this is done, the engineer must then adjust, or tune, the controller so that it minimizes the trial-and-error operation required to control. Most controllers have up to three terms used to tune them. To do a creditable job, the engineer must first know the characteristics of the process to be controlled. Once these characteristics are known, the control system can be designed, and the controller can be tuned. What is meant by process characteristics

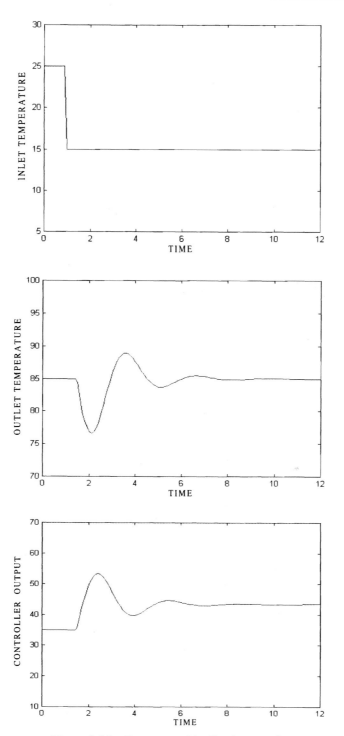

Figure 1-5.1   Response of feedback control.

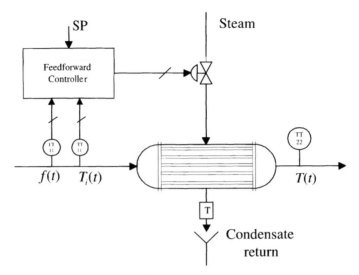

Figure 1-5.2 Feedforward control.

is explained in Chapter 2; in Chapter 3 we present various methods to tune controllers.

1-5.2 Feedforward Control

Feedback control is the most common control strategy in the process industries. Its simplicity accounts for its popularity. In some processes, however, feedback control may not provide the control performance required. For these processes, other types of control may have to be designed. In Chapters 5 and 7 we present additional control strategies that have proven to be profitable. One such strategy is *feedforward control*. The objective of feedforward control is to measure the disturbances and compensate for them before the controlled variable deviates from the set point. If applied correctly, the controlled variable deviation would be minimum.

A concrete example of feedforward control is the heat exchanger shown in Fig. 1-1.2. Suppose that "major" disturbances are the inlet temperature $T_i(t)$ and the process flow $f(t)$. To implement feedforward control these two disturbances must first be measured and then a decision made as to how to manipulate the steam valve to compensate for them. Figure 1-5.2 shows this control strategy. The feedforward controller makes the decision about how to manipulate the steam valve to maintain the controlled variable at set point, depending on the inlet temperature and process flow.

In Section 1-2 we learned that there are a number of different disturbances. The feedforward control system shown in Fig. 1-5.2 compensates for only two of them. If any of the other disturbances enter the process, this strategy will not compensate for it, and the result will be a permanent deviation from set point of the controlled variable. To avoid this deviation, some feedback compensation must be added to feedforward control; this is shown in Fig. 1-5.3. Feedforward control now compen-

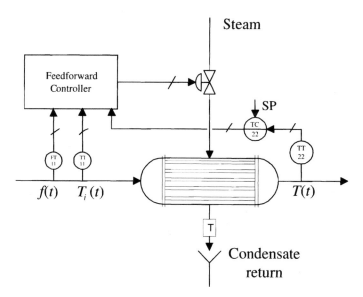

Figure 1-5.3   Feedforward control with feedback compensation.

sates for the "major" disturbances; feedback control compensates for all other disturbances. In Chapter 7 we present the development of the feedforward controller. Actual industrial cases are used to discuss this important strategy in detail.

It is important to notice that the three basic operations, M, D, A, are still present in this more "advanced" control strategy. The sensors and transmitters perform the measurement. Both feedforward and feedback controllers make the decision; the steam valve takes action.

The advanced control strategies are usually more costly, in hardware, computing power, and personnel necessary to design, implement, and maintain, than feedback control. Therefore, they must be justified (safety or economics) before they can be implemented. The best procedure is first to design and implement a simple control strategy, keeping in mind that if it does not prove satisfactory, a more advanced strategy may be justifiable. It is important, however, to recognize that these advanced strategies still require feedback compensation.

1-6 SUMMARY

In this chapter the need for automatic process control has been discussed. Industrial processes are not static but rather, very dynamic; they are changing continuously because of many types of disturbances. It is principally because of this dynamic nature that control systems are needed on a continuous and automatic basis to watch over the variables that must be controlled.

The working principles of a control system can be summarized with the three letters M, D, and A: M refers to the measurement of process variables, D to the decision to be made based on the measurements of the process variables, and A to the action to be taken based on the decision.

The basic components of a process control system were also presented: sensor/transmitter, controller, and final control element. The most common types of signals—pneumatic, electrical, and digital—were introduced along with the purpose of transducers.

Two control strategies were presented: feedback and feedforward control. The advantages and disadvantages of both strategies were discussed briefly.

CHAPTER 2

PROCESS CHARACTERISTICS

In this chapter we discuss process characteristics and describe in detail what is meant by a process, their characteristics, and how to obtain these characteristics using simple process information. The chapter is most important in the study of process control. Everything presented in this chapter is used to tune controllers and to design various control strategies.

2-1 PROCESS AND IMPORTANCE OF PROCESS CHARACTERISTICS

It is important at this time to describe what a process is from a controls point of view. To do this, consider the heat exchanger of Chapter 1, which is shown again in Fig. 2-1.1a. The controller's job is to control the process. In the example at hand, the controller is to control the outlet temperature. However, realize that the controller only receives the signal from the transmitter. It is through the transmitter that the controller "sees" the controlled variable. *Thus, as far as the controller is concerned, the controlled variable is the transmitter output.* The controller only looks at the process through the transmitter. The relation between the transmitter output and the process variable is given by the transmitter calibration.

In this example the controller is to manipulate the steam valve position to maintain the controlled variable at the set point. Realize, however, that the way the controller manipulates the valve position is by changing its signal to the valve (or transducer). Thus the controller does not manipulate the valve position directly; it only manipulates its output signal. Thus, *as far as the controller is concerned, the manipulated variable is its own output.*

If the controller is to control the process, we can therefore define the process as anything between the controller's output and the signal the controller receives. Referring to Fig. 2-1.1a, the process is anything within the area delineated by the curve. The process includes the I/P transducer, valve, heat exchanger with

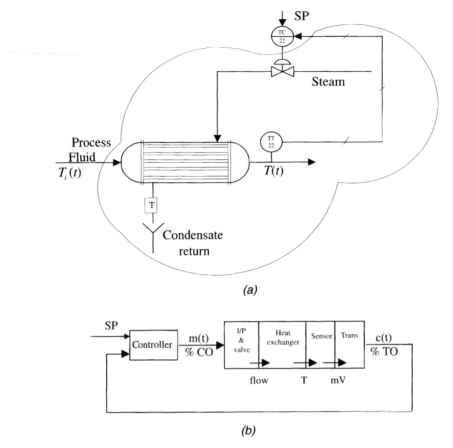

(a)

(b)

Figure 2-1.1 Heat exchanger temperature control system.

associated piping, sensor, and transmitter. That is, *the process is everything except the controller.*

A useful diagram is shown in Fig. 2-1.1*b*. The diagram shows all the parts of the process and how they relate. The diagram also clearly shows that the process output is the transmitter output and the process input is provided by the controller output. Note that we refer to the output of the transmitter *as* $c(t)$ to stress the fact that this signal is the real controlled variable; the unit of $c(t)$ is %TO (transmitter output). We refer to the signal from the controller as $m(t)$ to stress the fact that this signal is the real manipulated variable; the unit of $m(t)$ is %CO (controller output).

Now that we have defined the process to be controlled, it is necessary to explain why it is important to understand the terms that describe its characteristics. As we learned in Chapter 1, the control response depends on the tuning of the controller. The optimum tunings depend on the process to be controlled. As we well know, every process is different, and consequently, to tune the controller, the process characteristics must first be obtained. *What we do is to adapt the controller to the process.*

Another way to say that every process has different characteristics is to say that every process has its own "personality." If the controller is to provide good control, the controller personality (tunings) must be adapted to that of the process. It is important to realize that once a process is built and installed, it is not easy to change it. That is, the process is not very flexible. All the flexibility resides in the controller since it is very easy to change its tunings. As we show in Chapter 3, once the terms describing the process characteristics are known, the tuning of the controller is a rather simple procedure. Here lies the importance of obtaining the process characteristics.

2-2   TYPES OF PROCESSES

Processes can be classified into two general types depending on how they respond to an input change: self-regulating and non-self-regulating. The response of a *self-regulating process* to *step change* in input is depicted in Fig. 2-2.1. As shown in the

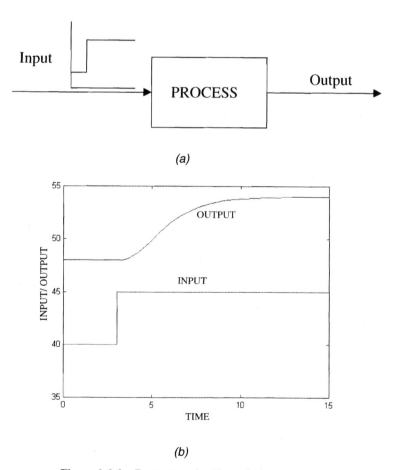

Figure 2-2.1   Response of self-regulating processes.

figure, upon a bound change in input, the output reaches a new final operating condition and remains there. That is, the process regulates itself to a new operating condition.

The response of *non-self-regulating processes* to a step change in input is shown in Fig. 2-2.2. The figure shows that upon a bound change in input, the process output does not reach, in principle, a final operating condition. That is, the process does not regulate itself to a new operating condition. The final condition will be an extreme operating condition, as we shall see.

Figure 2-2.2 shows two different responses. Figure 2-2.2*a* shows the output reaching a constant rate of change (slope). The typical example of this type of process is the level in a tank, as shown in Fig. 2-2.3. As the signal to the pump (process input) is reduced, the level in the tank (process output) starts to increase and reaches a steady rate of change. The final operating condition is when the tank overflows (extreme operating condition). Processes with this type of response are referred to as *integrating processes*. Not all level processes are of the integrating type, but they are the most common examples.

Figure 2-2.2*b* shows a response that changes exponentially. The typical example of this type of process is a reactor (Fig. 2-2.4) where an exothermic chemical reaction takes place. Suppose that the cooling capacity is somewhat reduced by closing the cooling valve (increasing the signal to the valve). Figure 2-2.2*b* shows that as the signal to the cooling valve (process input) increases, the water flow is reduced and the temperature in the reactor (process output) increases exponentially. The final operating condition is when the reactor melts down or when an explosion or any other extreme operating condition occurs (open a relief valve). This type of process is referred to as *open-loop unstable*. Certainly, the control of this type of process is quite critical. Not all exothermic chemical reactors are open-loop unstable, but they are the most common examples.

Sometimes the input variable is also referred to as a *forcing function*. This is so because it forces the process to respond. The output variable is sometimes referred to as a *responding variable* because it responds to the forcing function.

Fortunately, most processes are of the self-regulating type. In this chapter we discuss only this type. In Chapter 3 we present the method to tune level loops (integrating process).

2-3 SELF-REGULATING PROCESSES

There are two types of self-regulating processes: single capacitance and multi-capacitance.

2-3.1 Single-Capacitance Processes

The following two examples explain what it is meant by single-capacitance processes.

Example 2-3.1. Figure 2-3.1 shows a tank where a process stream is brought in, mixing occurs, and a stream flows out. We are interested in how the outlet temperature responds to a change in inlet temperature. Figure 2-3.2 shows how the outlet

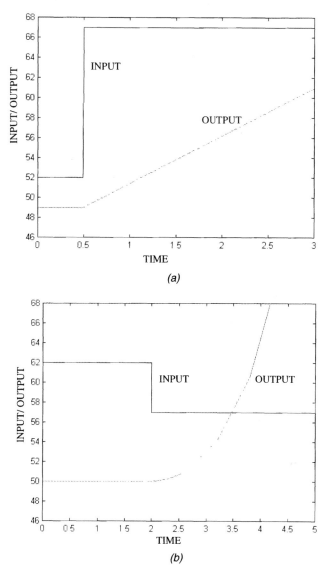

Figure 2-2.2   Response of non-self-regulating processes.

temperature responds to a step change in inlet temperature. The response curve shows the steepest slope occurring at the beginning of the response. This response to a step change in input is typical of all single-capacitance processes. Furthermore, this is the simplest way to recognize if a process is of single capacitance.

Example 2-3.2. Consider the gas tank shown in Fig. 2-3.3. Under steady-state conditions the outlet and inlet flows are equal and the pressure in the tank is constant. We are interested in how the pressure in the tank responds to a change in inlet flow,

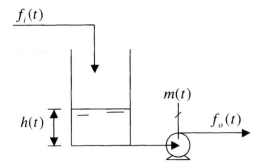

Figure 2-2.3   Liquid level.

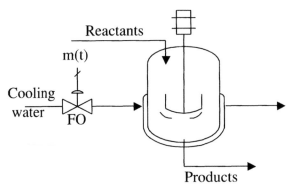

Figure 2-2.4   Chemical reactor.

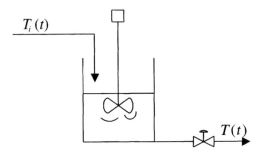

Figure 2-3.1   Process tank.

shown in Fig. 2-3.4*a*, and to a change in valve position, vp(*t*), shown in Fig. 2-3.4*b*. When the inlet flow increases, in a step change, the pressure in the tank also increases and reaches a new steady value. The response curve shows the steepest slope at the beginning. Consequently, the relation between the pressure in the tank and the inlet flow is that of a single capacitance. Figure 2-3.4*b* shows that when the outlet valve opens, the percent valve position increases in a step change, the pressure in the tank drops. The steepest slope on the response curve occurs at its

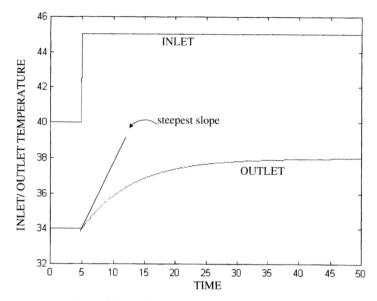

Figure 2-3.2   Response of outlet temperature.

Figure 2-3.3   Gas tank.

beginning, and therefore the relation between the pressure in the tank and the valve position is also of single capacitance.

Terms That Describe the Process Characteristics. We have so far shown two examples of single-capacitance processes. It is important now to define the terms that describe the characteristics of these processes; there are three such terms.

Process Gain (K). Process gain (or simply, *gain*) is defined as the ratio of the change in output, or responding variable, to the change in input, or forcing function. Mathematically, this is written

$$K = \frac{\Delta \text{Output}}{\Delta \text{Input}} = \frac{\Delta \text{Responding variable}}{\Delta \text{Forcing function}} = \frac{O_{\text{final}} - O_{\text{initial}}}{I_{\text{final}} - I_{\text{initial}}} \qquad (2\text{-}3.1)$$

Let us apply this definition of gain to Examples 2-3.1 and 2-3.2.
   For the thermal system, from Fig. 2-3.2, the gain is

$$K = \frac{\Delta T}{\Delta T_i} = \frac{(33 - 25)°\text{F outlet temperature}}{(35 - 25)°\text{F inlet temperature}} = 0.8 \frac{°\text{F outlet temperature}}{°\text{F inlet temperature}}$$

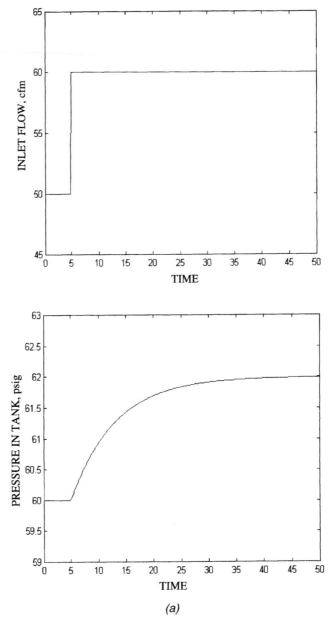

(a)

Figure 2-3.4 Response of pressure in tank to a change in (*a*) inlet flow and (*b*) valve position.

Therefore, the gain tells us how much the outlet temperature changes per unit change in inlet temperature. Specifically, it tells us that for a 1°F increase in inlet temperature, there is a 0.8°F increase in outlet temperature. Thus, this gain tells us how *sensitive* the outlet temperature is to a change in inlet temperature.

For the gas tank, from Fig. 2-3.4*a*, the gain is

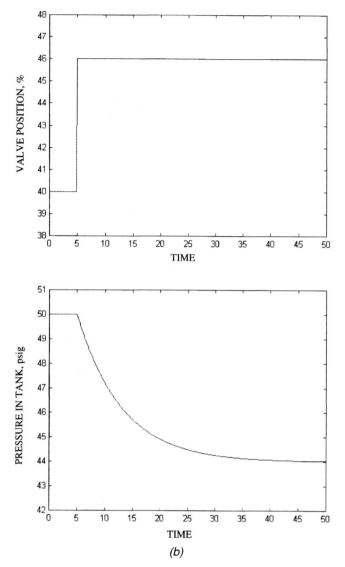

(b)

Figure 2-3.4 *Continued.*

$$K = \frac{\Delta p}{\Delta f_i} = \frac{(62-60)\text{psi}}{(60-50)\text{cfm}} = 0.2\,\frac{\text{psi}}{\text{cfm}}$$

This gain tells us how much the pressure in the tank changes per unit change in inlet flow. Specifically, it tells us that for a 1-cfm increase in inlet flow there is a 0.2-psi increase in pressure in the tank. As in the earlier example, the gain tells us the *sensitivity* of the output variable to a change in input variable.

Also for the gas tank, from Fig. 2-3.4*b*, another gain is

$$K = \frac{\Delta p}{\Delta vp} = \frac{(44 - 50)\text{psi}}{(46 - 40)\% \text{vp}} = -1.0 \frac{\text{psi}}{\% \text{vp}}$$

This gain tells us that for an increase of 1% in valve position the pressure in the tank decreases by 1.0 psi.

These examples indicate that the *process gain* (K) describes the *sensitivity* of the output variable to a change in input variable. The output could be the controlled variable and the input, the manipulated variable. Thus, in this case, the gain then describes how sensitive the controlled variable is to a change in the manipulated variable.

Anytime the process gain is specified, three things must be given:

1. *Sign.* A positive sign indicates that if the process input increases, the process output also increases; that is, both variables move in the same direction. A negative sign indicates the opposite; that is, the process input and process output move in the opposite direction. Figure 2-3.4b shows an example of this negative gain.
2. *Numerical value.*
3. *Units.* In every process these are different types of gains. Consider the gas tank example. Figure 2-3.4a provides the gain relating the pressure in the tank to the inlet flow and consequently, the unit is psi/cfm. Figure 2-3.4b provides the gain relating the pressure in the tank to the valve position, and consequently, the unit is psi/%vp. If the sign and numerical value of the gain are given, the only thing that would specify what two variables are related by a particular gain are the units. In every process there are many different variables and thus different gains.

It is important to realize that the gain relates only steady-state values, that is, how much a change in the input variable affects the output variable. Therefore, *the gain is a steady-state characteristic of the process.* The gain does not tell us anything about the dynamics of the process, that is, how fast changes occur.

To describe the dynamics of the process, the following two terms are needed: the time constant and the dead time.

Process Time Constant (τ). The process time constant (or simply, *time constant*) for a single-capacitance processes is defined [1], from theory, as

τ = Amount of time counted from the moment the variable starts to respond that it takes the process variable to reach 63.2% of its total change

Figure 2-3.5, a duplicate of Fig. 2-3.4b, indicates the time constant. It is seen from this figure, and therefore from its definition, that the time constant is related to the speed of response of the process. The faster a process responds to an input, the shorter the time constant; the slower the process responds, the longer the time constant. The process reaches 99.3% of the total change in 5τ from the moment it starts to respond, or in 99.8% in 6τ. The unit of time constant is minutes or seconds. The unit used should be consistent with the time unit used by the controller or control

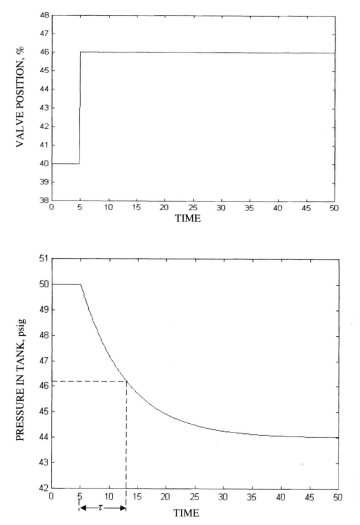

Figure 2-3.5   Response of pressure in tank to a change in valve position, time constant.

system. As discussed in Chapter 3, most controllers use minutes as time units, while a few others use seconds.

To summarize, the time constant tells us how fast a process responds once it starts to respond to an input. Thus, the time constant is a term related to the dynamics of the process.

Process Dead Time (t_o).   Figure 2-3.6 shows the meaning of process dead time (or simply, *dead time*). The figure shows that

t_o = finite amount of time between the change in input variable
and when the output variable starts to respond

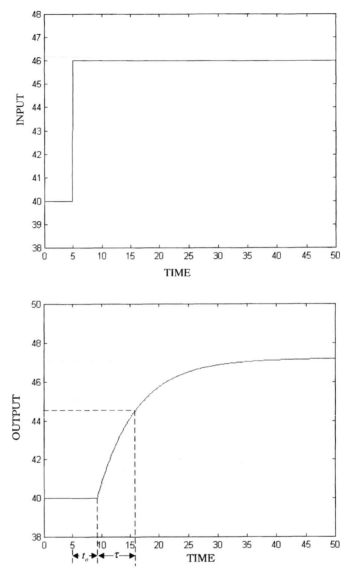

Figure 2-3.6   Meaning of dead time.

The figure also shows the time constant to aid in understanding the difference between them. Both τ and t_o are related to the dynamics of the process.

As we will learn shortly, most processes have some amount of dead time. *Dead time has significant adverse effects on the controllability of control systems.* This is shown in detail in Chapter 5.

The numerical values of K, τ, and t_o depend on the physical parameters of the process. That is, the numerical values of K, τ, and t_o depend on the size, calibration,

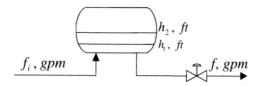

Figure 2-3.7   Horizontal tank with dished ends.

and other physical parameters of the equipment and process. If any of these changes, the process will change and this will be reflected in a change in K, τ, and t_o; the terms will change singly or in any combination.

Process Nonlinearities. *The numerical value of K, τ, and t_o depend on the process operating conditions.* Processes where the numerical values of K, τ, and t_o are constant over the entire operating range, known as *linear processes*, occur very infrequently. Most often, processes are *nonlinear*. In these processes the numerical values of K, τ, and t_o vary with operating conditions. Nonlinear processes are the norm.

Figure 2-3.7 shows a simple example of a nonlinear process. A horizontal tank with dished ends is shown with two different heights, h_1 and h_2. Because the cross section of the tank at h_1 is less than at h_2, the level at h_1 will respond faster to changes in inlet, or outlet, flow than the level at h_2. That is, the dynamics of the process at h_1 are faster than at h_2. A detailed analysis of the process shows that the gain depends on the square root of the pressure drop across the valve. This pressure drop depends on the liquid head in the tank. Thus the numerical value of the gain will vary as the liquid head in the tank varies.

The tank process is mainly nonlinear because of the shape of the tank. Most processes are nonlinear, however, because of their physical–chemical characteristics. To mention a few, consider the relation between the temperature and the rate of reaction (exponential, the Arrhenius expression); between the temperature and the vapor pressure (another exponential, the Antoine expression); between flow through a pipe and the heat transfer coefficients; and finally, the pH.

The nonlinear characteristics of processes are most important from a process control point of view. As we have already discussed, the controller is always adapted to the process. Thus, if the process characteristics change with operating conditions, the controller tunings should also change, to maintain control performance.

Mathematical Description of Single-Capacitance Processes. Mathematics provides the technical person with a very convenient communication tool. The equation that describes how the output variable, $O(t)$, of a single-capacitance process, with no dead time, responds to a change in input variable, $I(t)$, is given by the differential equation

$$\tau \frac{dO(t)}{dI(t)} + O(t) = KI(t) \tag{2-3.2}$$

We do not usually use differential equations in process control studies, but rather, transform them into the shorthand form

$$\frac{O(s)}{I(s)} = \frac{K}{\tau s + 1} \tag{2-3.3}$$

This equation is referred to as a *transfer function* because it describes how the process "transfers" the input variable to the output variable. Some readers may remember that the s term refers to the Laplace operator. For those readers that may not have seen it before, don't worry: s stands for "shorthand." We will only use this equation to describe single-capacitance processes, not to do any mathematics. Equation (2-3.3) develops from Eq. (2-3.2), and because this equation is a first-order differential equation, single-capacitance processes are also called *first-order processes*.

The transfer function for a pure dead time is given by the transfer function

$$\frac{O(s)}{I(s)} = e^{-t_o s} \tag{2-3.4}$$

Thus, the transfer function for a first-order-plus-dead-time (FOPDT) process is given by

$$\frac{O(s)}{I(s)} = \frac{K e^{-t_o s}}{\tau s + 1} \tag{2-3.5}$$

Transfer functions will be used in these notes to facilitate communication and to describe processes.

2-3.2 Multicapacitance Processes

The following two examples explain the meaning of multicapacitance.

Example 2-3.3. Consider the tanks-in-series process shown in Fig. 2-3.8. This process is an extension of the single tank shown in Fig. 2-3.1. We are interested in learning how the outlet temperature from each tank responds to a step change in inlet temperature to the first tank, $T_i(t)$; each tank is assumed to be well mixed. Figure 2-3.8 also shows the response curves. The response curve of $T_1(t)$ shows the first tank behaving as a first-order process. Thus its transfer function is given by

$$\frac{T_1(s)}{T_i(s)} = \frac{K_1}{\tau_1 s + 1} \tag{2-3.6}$$

The $T_2(t)$ curve shows the steepest slope occurring later in the curve, not at the beginning of the response. What happens is that once $T_i(t)$ changes, $T_1(t)$ has to change enough before $T_2(t)$ starts to change. Thus, at the very beginning, $T_2(t)$ is barely changing. When the process is composed of the first two tanks, it is not of first order. Since we know there are two tanks in series in this process, we write its transfer function as

$$\frac{T_2(s)}{T_i(s)} = \frac{K_2}{(\tau_1 s + 1)(\tau_2 s + 1)} \tag{2-3.7}$$

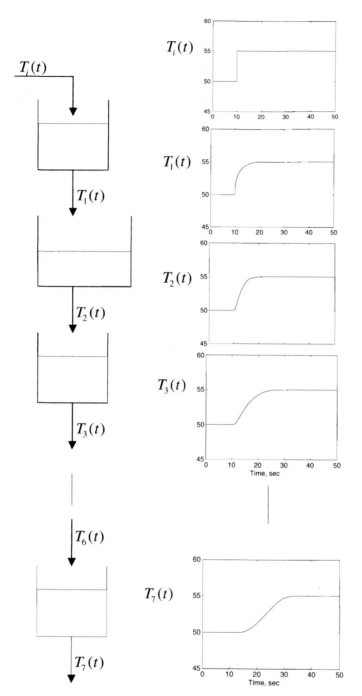

Figure 2-3.8   Tanks in series.

The $T_3(t)$ curve shows an even slower response than before. $T_3(t)$ has to wait now for $T_2(t)$ to change enough before it starts to respond. Since there are three tanks in series in this process, we write its transfer function as

$$\frac{T_3(s)}{T_i(s)} = \frac{K_3}{(\tau_1 s + 1)(\tau_2 s + 1)(\tau_3 s + 1)} \tag{2-3.8}$$

All of our previous comments can be extended when we consider four tanks as a process. In this case we write the transfer function

$$\frac{T_4(s)}{T_i(s)} = \frac{K_4}{(\tau_1 s + 1)(\tau_2 s + 1)(\tau_3 s + 1)(\tau_4 s + 1)} \tag{2-3.9}$$

and so on.

Since a process described by Eq. (2-3.6) is referred to as a first-order process, we could refer to a process described by Eq. (2-3.7) as a second-order process. Similarly, the process described by Eq. (2-3.8) is referred to as a third-order process, the process described by Eq. (2-3.9) as a fourth-order process, and so on. In practice when a curve such as the one given by $T_2(t)$, $T_3(t)$, or $T_4(t)$ is obtained, we really do not know the order of the process. Therefore, any process that is not of first order is referred to as a *higher-order* or *multicapacitance process*.

Figure 2-3.8 shows that as the order of the process increases, the response looks as if it has dead time. As a matter of fact, this "apparent," or "effective," dead time increases as the order of the process increases. Since most processes are of a higher order, this is a common reason why dead times are found in processes.

To avoid dealing with multiple time constants, second-order-plus-dead-time (SOPDT) or first-order-plus-dead-time (FOPDT) approximations to higher-order processes are commonly used:

$$\frac{O(s)}{I(s)} = \frac{K}{\prod_{i=1}^{n}(\tau_i s + 1)} \approx \frac{Ke^{-t_0 s}}{(\tau_1 s + 1)(\tau_2 s + 1)} \tag{2-3.10}$$

or

$$\frac{O(s)}{I(s)} = \frac{K}{\prod_{i=1}^{n}(\tau_i s + 1)} \approx \frac{Ke^{-t_0 s}}{\tau s + 1} \tag{2-3.11}$$

We show how to obtain these approximations in Section 2-5.

Example 2-3.4. As a second example of a multicapacitance process, consider the reactor shown in Fig. 2-3.9. The well-known exothermic reaction A → B occurs in this reactor; a cooling jacket surrounds the reactor to remove the heat of reaction. A thermocouple inside a thermowell is used to measure the temperature in the reactor. It is desired to know how the process temperatures change if the inlet temperature to the jacket, $T_{J_i}(t)$, changes; the responses are also shown in Fig. 2-3.9. The figure shows that once $T_{J_i}(t)$ changes, the first variable that responds is the jacket

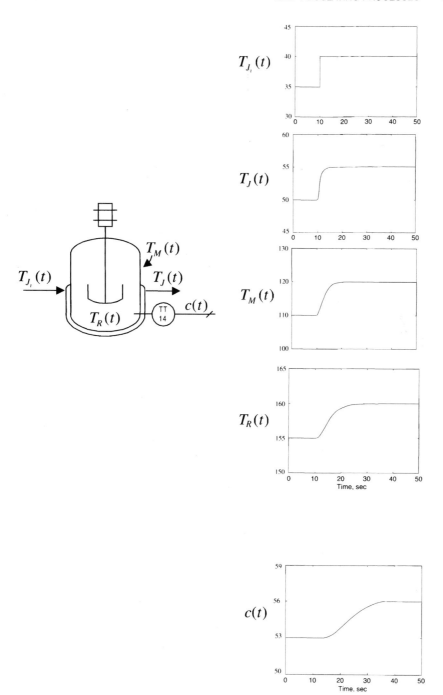

Figure 2-3.9   Exothermic chemical reactor.

temperature, $T_J(t)$. For this example we have assumed the jacket to be well mixed, and thus the temperature responds as a first-order process. The second variable that responds is the temperature of the metal wall, $T_M(t)$. The amount that $T_M(t)$ changes depends on the volume of the metal, density of the metal, heat capacity of the metal, and so on. Also, how fast $T_M(t)$ changes depend on the thickness of the wall, thermal conductivity of the metal, and so on. That is, the process characteristics depend on the physical parameters (material of construction and sizes) of the process. This is what we discussed in Section 2-3.1. The temperature in the reactor, $T_R(t)$, responds next. Finally, the output signal from the sensor/transmitter, $c(t)$, starts to react. How fast this signal changes depends on whether the thermocouple (sensor) is a bare thermocouple or if it is inside a thermowell.

The important thing to learn from this example is that every time a capacitance is encountered, it slows the dynamics (longer τ and t_o) of the process.

To summarize, multicapacitance, or higher-order processes, are most often encountered. The reason for this is that processes are usually formed by single capacitances in series.

2-4   TRANSMITTERS AND OTHER ACCESSORIES

Let us look at the characteristics of transmitters and transducers. Consider an electronic (4 to 20 mA) pressure transmitter with a calibration of 0 to 50 psig process pressure. To calculate the gain of this transmitter, we follow the definition of gain:

$$K_T = \frac{\Delta O}{\Delta I} = \frac{(20-4)\text{mA TO}}{(50-0)\text{psi}} = 0.32 \frac{\text{mA TO}}{\text{psi}}$$

or

$$K_T = \frac{\Delta O}{\Delta I} = \frac{(100-0)\% \text{TO}}{(50-0)\text{psi}} = 2 \frac{\% \text{TO}}{\text{psi}}$$

depending on whether the output from the transmitter is considered in mA or in percent.

The dynamics (τ and t_o) of sensor/transmitters are often, although not always, fast compared to the process unit. The method we learn to characterize the process will be such that it considers the dynamics of the valve, process unit, and sensor/transmitter all together, as one. Thus, there is no need to discuss in detail the dynamics of sensors/transmitters. There are, however, some units, such as chromatographs, that may add significant dead time to the process. As mentioned earlier, dead time has a significantly adverse effect on the controllability of processes.

Consider a current-to-pneumatic (I/P) transducer. The gain of this transducer is

$$K_T = \frac{\Delta O}{\Delta I} = \frac{(15-3)\text{psi}}{(20-4)\text{mA}} = 0.75 \frac{\text{psi}}{\text{mA}}$$

or

$$K_T = \frac{\Delta O}{\Delta I} = \frac{(100-0)\%\,\text{output}}{(100-0)\%\,\text{input}} = 1.0\frac{\%\,\text{output}}{\%\,\text{input}}$$

depending on the units desired.

2-5 OBTAINING PROCESS CHARACTERISTICS FROM PROCESS DATA

In this section we learn how to obtain the process characteristics (K, τ, and t_o) from process data for self-regulating processes. We have already learned that most processes are self-regulating and of higher order, with a general transfer function as

$$\frac{O(s)}{I(s)} = \frac{K}{(\tau_1 s + 1)(\tau_2 s + 1)\cdots(\tau_n s + 1)} \tag{2-5.1}$$

As mentioned earlier, however, higher-order processes can be approximated by a second-order-plus-dead-time (SOPDT) transfer function, Eq. (2-3.10). What happens in practice, though, is that there is no easy, reliable, and consistent method to approximate a higher-order process by this type of transfer function. What it is usually done, therefore, is to approximate a higher-order system by a first-order-plus-dead-time (FOPDT) transfer function, Eq. (2-3.11). Thus we approximate higher-order processes by a low-order-plus-dead-time model. The method presented next is the most objective of all those available, the one that gives the best approximation, and the easiest one to use. (The day this method was developed, Murphy was sleeping!)

To use a concrete example, consider the heat exchanger shown in Fig. 2-1.1a. Assume that the temperature transmitter has a calibration of 100 to 250°C. To obtain the necessary process data, the following steps are used:

1. Set the controller to manual mode. Effectively, the controller is removed.
2. Make a step change in the controller output.
3. Record the process variable.

For example, suppose that these steps are followed in the heat exchanger example and the results are those shown in Fig. 2-5.1. The response curve indicates that this exchanger is a higher-order process.

To obtain the dynamic terms τ and t_o, we make use of the *two-point method* (or fit 3 in Ref. 1). The method consists in obtaining two data points from the response curve (*process reaction curve*). These two points are the time it takes the process to reach 63.2% of the total change in output, or $t_{0.632\Delta O}$, and the time it takes the process to reach 28.3% of the total change in output, or $t_{0.283\Delta O}$; these two points are shown in Fig. 2-5.1. Time zero is the time when the step change in controller output occurs. With these two data points, τ and t_o are obtained from the following equations:

$$\tau = 1.5(t_{0.632\Delta O} - t_{0.283\Delta O}) \tag{2-5.2}$$

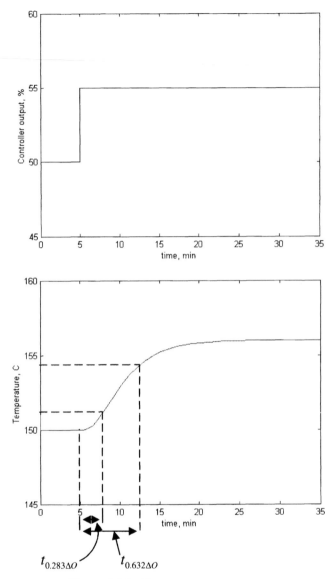

Figure 2-5.1   Process response curve.

$$t_o = t_{0.632\Delta O} - \tau \tag{2-5.3}$$

The units of τ and t_o are the same time units as those used by the control system.

Now that τ and t_o have been evaluated, we proceed to evaluate K. Following the definition of gain,

$$K = \frac{(156-150)\,°C}{(55-50)\,\%\,CO} = \frac{6\,°C}{5\,\%\,CO} = 1.2\frac{°C}{\%\,CO}$$

This gain says that at the present operating condition, a change of 1%CO results in a change of 1.2°C in outlet process temperature. This gain correctly describes the sensitivity of the outlet process temperature to a change in controller output. However, this gain is only a partial process gain and not the *total process gain*. The total process gain is the one that says how much the process output, $c(t)$ in %TO, changes per change in process input, $m(t)$ in %CO; the reader may refer to Fig. 2-1.1b to understand this point. That is, the process output is given by the transmitter output and it is not the temperature. Therefore, we are interested in how much the transmitter output changes per change in controller output, or

$$K = \frac{\Delta c}{\Delta m} = \frac{\text{change in transmiter's output, }\%\text{TO}}{\text{change in controller's output, }\%\text{CO}} \qquad (2\text{-}5.4)$$

The change in transmitter output is calculated as follows:

$$\Delta c = \left(\frac{6°C}{150°C} \right) 100\% = 4\%\text{TO}$$

or, in general,

$$\Delta c = \left(\frac{\Delta PV}{\text{span}} \right) 100\%\text{TO} = \left(\frac{\text{change in process variable in engineering units}}{\text{span of transmitter}} \right) 100\%\text{TO}$$

$$(2\text{-}5.5)$$

Therefore, the total process gain for this example is

$$K = \frac{4\%\text{TO}}{5\%\text{CO}} = 0.8 \frac{\%\text{TO}}{\%\text{CO}}$$

If the process variable had been recorded in percent of transmitter output, there would be no need for any extra calculation.

We can now write the transfer function for this process as

$$\frac{C(s)}{M(s)} = \frac{0.8 e^{-t_o s}}{\tau s + 1}$$

This transfer function describes the relation between the transmitter output and the controller output. If a transfer function describing the relation between the transmitter output and any other process input (other than the controller output) is desired, the same procedure is then followed to evaluate K, τ, and t_o. In this case the units of the K will be different than before; that is, they will not be %TO/%CO. The units will depend on the units of the particular input.

To illustrate the approximation using the two-point method, consider a third-order process described by the following transfer function:

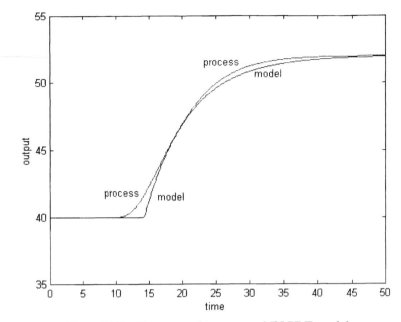

Figure 2-5.2   Response of process and FOPDT model.

$$\frac{O(s)}{I(s)} = \frac{1.2}{(3s+1)(3s+1)(4s+1)}$$

Suppose that the input $I(t)$ is changed by 10 input units at time 10 time units. The output $O(t)$ is recorded, and the mathematical calculations described above are followed. The FOPDT model calculated is

$$\frac{O(s)}{I(s)} = \frac{1.2e^{-4.25s}}{6.75s+1}$$

Figure 2-5.2 shows the responses of the process and of the model to the same step change in input. The responses are quite close. Note that the model shows a longer dead time than the "apparent dead time" in the process. This will always be the case, and it is done in an effort to minimize the area between the two curves.

As mentioned earlier, the procedure just presented provides the "best" approximation of a higher-order (or multicapacitance) process by a first-order-plus-dead-time (or single-capacitance) process. It provides an important tool to process control personnel.

2-6   QUESTIONS WHEN PERFORMING PROCESS TESTING

The following questions must be answered when performing process testing.

1. In what direction should the controller output be moved?

Safety is the most important consideration. You always want to go in a safe direction.

2. By how much (%) should the controller output be moved?

Move by the smallest amount that gives you a good (readable) answer. There are two reasons for this: (1) if you move the process far from its present steady state, this makes the operating personnel nervous; and (2) you want to obtain the characteristics close to the operating point, because much away from it, nonlinearities may start having an effect.

3. How many tests should be performed?

We want to have repeatability in the results; therefore, we could say that we should have as many tests as possible to ensure repeatability. However, many tests are not realistic either. To start, perhaps two tests, in different directions, are enough. Once the numerical values of the characteristics are obtained, they can be compared, and if not similar, more tests may be justified. If they are similar, an average can then be calculated.

4. What about noise?

Noise is a fact in many processes. Once a recording is obtained, an average process curve can be drawn freehand to obtain an average curve. This in itself is a way to filter the noise.

5. What if an upset enters the process while it is being tested?

The purpose is to learn how the manipulated variable affects the controlled variable. If a disturbance enters the process while it is being tested, the results will be due to both inputs (manipulated variable and disturbance). It is very difficult to deconvolute the effects. This disturbance may be the reason why two tests may provide much different results. The test should be done under the most possible steady-state conditions.

2-7 SUMMARY

In this chapter we have only considered the process. We have looked at the characteristics, or personality, of processes. Specifically, we defined and discussed the following terms: process, self-regulating and non-self-regulating processes, integrating and open-loop unstable processes, single- and multicapacitance processes, process gain, process time constant, process dead time, process nonlinearities, and transfer functions. Finally, we presented and discussed a method to evaluate the process characteristics empirically.

What we learned in this chapter is most useful in tuning controllers (Chapters 3 and 4), in evaluating control system stability (Chapter 6), and in designing advanced control strategies (Chapters 7 to 9).

REFERENCE

1. C. A. Smith and A. B. Corripio, *Principles and Practice of Automatic Process Control*, 2nd ed., Wiley, New York, 1997.

PROBLEMS

2-1. Consider the bottoms of the separation column shown in Fig. P2-1*a*. In this column the temperature in a tray is controlled, manipulating the steam valve to the reboiler. The temperature transmitter has a range of 60 to 120°C. The controller was set in manual and its output was changed by +5% at time = 2 min. Fig. P2-1*b* shows the temperature response. Calculate the process characteristics K, τ, and t_o.

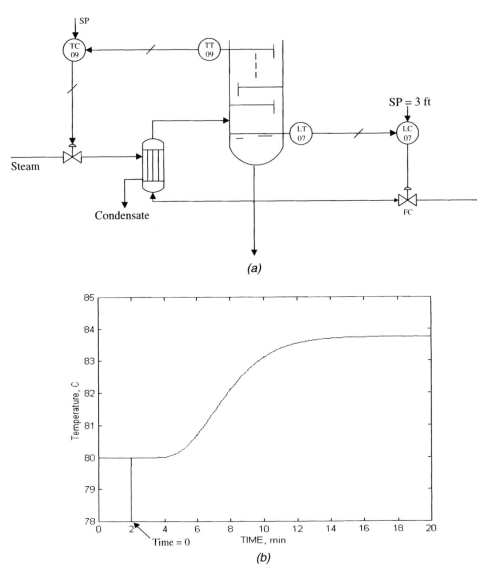

(a)

(b)

Figure P2-1   (*a*) Separation column; (*b*) response of temperature in separation column.

2-2. Consider the control system for the mixer in Fig. P2-2a. A very concentrated reactant is somewhat diluted with water before entering a reactor. The dilution of the reactant must be controlled for safety considerations. The density of the diluted stream is easily measured, and it is a very good indication of the dilution. The density transmitter has a range of 50 to 60 lb/ft^3. The density controller was set in manual and its output was changed by –3% at time = 1 min, as shown in Fig. P2-2b. Obtain the process characteristics K, τ, and t_o for this process.

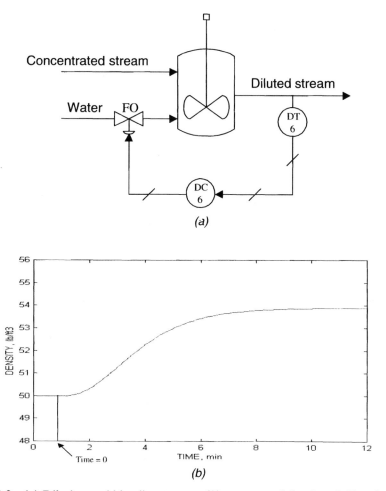

(a)

(b)

Figure P2-2   (a) Dilution and blending process; (b) response of density of diluted stream.

2-3. Consider the process to dry rock pellets (Fig. P2-3a). Wet pellets enter the drier to be dried before feeding them to a reactor. The moisture of the pellets exiting the drier must be controlled; the figure shows the control scheme. The moisture transmitter has a range of 2 to 6% moisture. It is desired to obtain the characteristics K, τ, and t_o of this process. To do so, the controller was set into manual and its output changed as shown in Fig. P2-3b; the figure also shows the process response. Find the process characteristics.

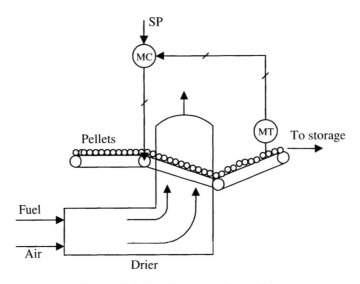

Figure P2-3 (a)   Process to dry pellets.

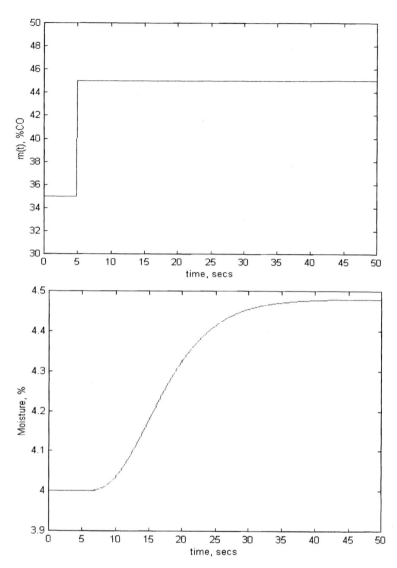

Figure P2-3 (*b*) Process response.

CHAPTER 3

FEEDBACK CONTROLLERS

In this chapter we present the most important types of industrial controllers. These controllers are used in analog systems, in distributed control systems (DCSs), and in stand-alone controllers, also sometimes referred to as single-loop controllers, or simply loop controllers. The DCSs and the stand-alone controllers are computer-based, and consequently, they do not process the signals on a continuous basis but rather, in a discrete fashion. However, the sampling time for these systems is rather fast, usually ranging from 10 times per second to about once per second. Thus, for all practical purposes, these controllers appear to be continuous.

3-1 ACTION OF CONTROLLERS

The selection of controller action is critical. *If the action is not selected correctly, the controller will not control.* Let us see how to select the action, and what it means.

Consider the heat exchanger control loop shown in Fig. 3-1.1. The process is at steady state and at the set point; the set point is constant. Assume that signal from the transmitter increases, indicating that the outlet temperature has increased above the set point. To return this temperature to the set point, the controller must close the steam valve by some amount. Because the valve is fail-closed (FC), the controller must reduce its output signal to close the valve (see the arrows in the figure). When an increase in signal to the controller requires a decrease in signal from the controller, the controller must be set to *reverse action*. Often, the term *increase/decrease*, or simply *decrease*, is also used. The *increase* refers to the measurement, $c(t)$, and the *decrease* refers to the manipulated variable, $m(t)$. The set point is not part of the decision.

Alternatively, consider the level loop shown in Fig. 3-1.2. The process is at steady state and at set point; the set point is constant. If the signal from the transmitter

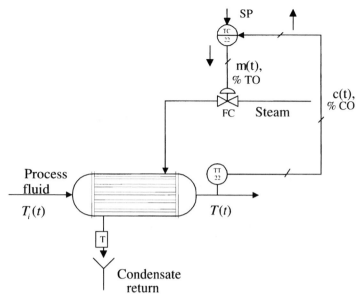

Figure 3-1.1   Heat exchanger control loop.

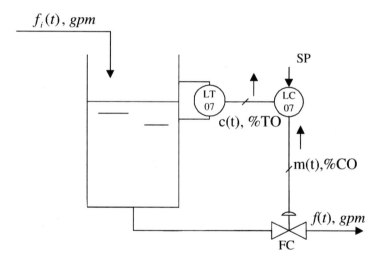

Figure 3-1.2   Liquid level control loop.

increases, indicating an increase in level, it is necessary to open the outlet valve more. Because the valve is fail-closed, the signal from the controller must increase to open the valve. Therefore, when the signal from the transmitter increases, the signal from the controller must also increase (see the arrows in the figure). In this case the controller must be set to *direct action*. Often, the term *increase/increase*, or simply in*crease*, is also used. The controller action is usually set by a switch, by a configuration bit, or by answering a question on most controllers.

3-2   TYPES OF FEEDBACK CONTROLLERS

The way that feedback controllers make a decision is by solving an equation based on the difference between the set point and the controlled variable. In this section we look at the most common types of controllers by looking at the equations that describe their operation.

As presented in Chapter 1, the signals entering and exiting the controllers are either electrical or pneumatic. Even in computer systems the signals entering from the field are electrical before they are converted by an analog-to-digital (A/D) converter to digital signals. Similarly, the signal the computer system sends back to the field is an electrical signal. To help simplify the presentation that follows, we will use all signals in percent, that is, 0 to 100% as opposed to 4 to 20mA or 3 to 15psig.

As mentioned, feedback controllers decide what to do to maintain the controlled variable at the set point by solving an equation based on the difference between the set point and the controlled variable. This difference, or error, is computed as

$$e(t) = c^{\text{set}}(t) - c(t) \tag{3-2.1}$$

where $c(t)$ is the controlled variable (most often, the controlled variable is given by the transmitter output and consequently, has units of %TO); $c^{\text{set}}(t)$ is the set point (this is the desired value of the controlled variable and thus has units of %TO; as the set point is entered in engineering units of the process variable, it is converted by the control system (controller) into %TO using the transmitter range); and $e(t)$ is the error in %TO. The error could have also being computed as $e(t) = c(t) - c^{\text{set}}(t)$; however, Eq. (3-2.1) will be the convention used in this book.

3-2.1   Proportional Controller

The proportional (P) controller is the simplest type of controller we will discuss. The equation that describes its operation is

$$m(t) = \overline{m} + K_C e(t) \tag{3-2.2}$$

where $m(t)$ is the controller output, %CO [the term $m(t)$ is used to stress that as far as the controller is concerned, this output is the manipulated variable]; K_C is the *controller gain*, %CO/%TO; and $\overline{m}$ is the bias value, %CO (this is the output from the controller when the error is zero; $\overline{m}$ is a constant value and it is also the output when the controller is switched to manual; it is very often initially set at midscale, 50%CO).

Equation (3-2.2) shows that the output of the controller is proportional to the error. The proportionality is given by the controller gain, K_C. The significance of this gain is shown graphically in Fig. 3-2.1. The figure shows that the larger K_C value, the more the controller output changes for a given error. Thus K_C *establishes the sensitivity of the controller to an error, that is, how much the controller output changes per unit change in error.* In other words, K_C establishes the aggressiveness of the controller. *The larger K_C is, the more aggressive the controller reacts to an error.*

Proportional controllers have the advantage of only one tuning parameter, K_C. However, they suffer a major disadvantage—that of operating the controlled

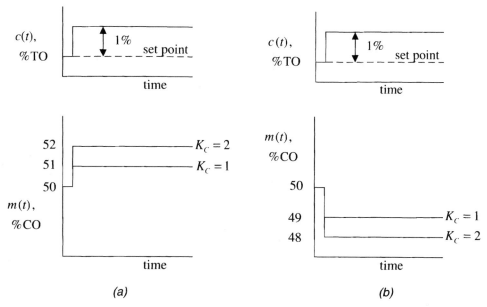

Figure 3-2.1   Effect of controller gain on output from controller: (*a*) direct-acting controller; (*b*) reverse-acting controller.

variable with an *offset*. Offset can be described as a *steady-state deviation of the controlled variable from the set point*, or just simply a *steady-state error*. To further explain the meaning of offset, consider the liquid-level control loop shown in Fig. 3-1.2. The design operating conditions are $f_i = f = 150$ gpm and $h = 4$ ft. Let us also assume that for the outlet valve to deliver 150 gpm, the signal to it must be 50%CO. If the inlet flow $f_i(t)$ increases, the response of the system with a proportional controller is shown in Fig. 3-2.2. The controller returns the controlled variable to a steady value but not to the set point required. The difference between the set point and the new steady state is the offset. The proportional controller is not "intelligent enough" to drive the controlled variable back to set point. The new steady-state value satisfies the controller.

Figure 3-2.2 shows four response curves corresponding to four different values of K_C. The figure shows that the larger the value of K_C, the smaller the offset. Therefore, one tends to ask why not set a maximum gain to eliminate the offset? Figure 3-2.2 also shows that while the larger K_C reduces the offset, the process response becomes more oscillatory. For most processes there is a maximum value of K_C beyond which the process goes unstable. This maximum value of K_C is called the *ultimate gain*, and we represent it as K_{C_U}. Thus there is a limit to the value of K_C that can be set in the controller and at the same time maintain stability. Therefore, the offset cannot be eliminated completely. Figure 3-2.2 shows that for the level loop of Fig. 3-1.2,

$$K_{C_U} \approx 1.55 \frac{\% \text{CO}}{\% \text{TO}}$$

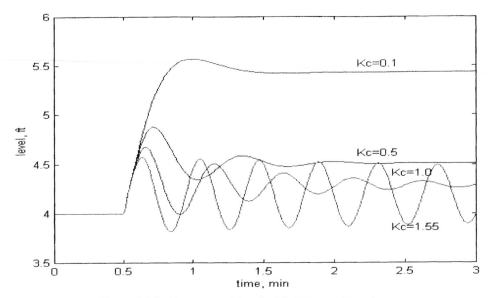

Figure 3-2.2   Response of level with different K_C values.

The obvious question is: Why does this offset occur? Let us now look at a simple explanation to this question. Consider again the liquid-level control system shown in Fig. 3-1.2 with the same operating conditions as those given previously, that is, $f_i = f = 150$ gpm, $h = 4$ ft, and with a required signal to the valve of 50% CO to deliver 150 gpm. Assume now that the inlet flow increases up to 170 gpm. When this happens the liquid level increases and the controller will in turn increase its output to open the valve to bring the level back down. To reach a steady operation, the outlet flow must now be 170 gpm. To deliver this new flow, the outlet valve must be open more than before, when it needed to deliver 150 gpm. Because the valve is fail-closed, let us assume that the new signal to the valve required to deliver 170 gpm is 60%. That is, the output from the controller must be 60%. Looking back at Eq. (3-2.2), we note that the only way for the controller output to be 60% is for the second term, on the right-hand side to have a value of +10% (remember, the bias value, $\bar{m}$, is set at mid-scale of 50%). To obtain this 10% from the second term, the error term must not be zero at steady state. *This required steady-state error is the offset.*

$$m(t) = 50\% + K_C e(t) = 50\% + 10\% = 60\%$$

Two points need to be stressed.

1. The magnitude of the offset depends on the value of the controller gain. Because the second term must have a value of +10% CO, the values are:

K_C	Offset, $e(t)$, (%TO)
1	10
2	5.0
4	2.5

As mentioned previously, the larger the gain, the smaller the offset. The reader must remember that above a certain K_C, most processes go unstable. However, the controller equation does not show this.

2. It seems that all a proportional controller is doing is reaching a steady state operating condition. Once a steady state is reached, the controller is satisfied. The amount of deviation from the set point, or offset, depends on the controller gain.

Many controller manufacturers do not use the term K_C for the tuning parameter; they use the term *proportional band*, PB. The relationship between gain and proportional band is given by

$$PB = \frac{100}{K_C} \tag{3-2.3}$$

In these cases the equation that describes the proportional controller is written as

$$m(t) = \overline{m} + \frac{100}{PB} e(t) \tag{3-2.4}$$

PB is usually referred to as *percent proportional band*.

Equation (3-2.3) presents a most important fact. A large controller gain is the same as a low, or narrow, proportional band, and a low controller gain is the same as a large, or wide, proportional band. An increase in PB is similar to a decrease in K_C, resulting in a less aggressive controller to an error. A decrease in PB is similar to an increase in K_C, resulting in a more aggressive controller. K_C and PB are the reciprocal of each other, and thus care must be taken when tuning the controller.

Let us offer another definition of proportional band. *Proportional band refers to the error (expressed in percentage of the range of the controlled variable) required to move the output of the controller from its lowest to its highest value.* Consider the heat exchanger control loop shown in Fig. 3-1.1. The temperature transmitter has a range from 100 to 300°C, and the set point of the controller is at 200°C. Figure 3-2.3 gives a graphical explanation of this definition of PB. The figure shows that a 100% PB means that as the controlled variable varies by 100% of its range, the controller output varies by 100% of its range. A 50% PB means that as the controlled variable varies by 50% of its range, the controller output varies by 100% of its range. Also notice that a proportional controller with a 200% PB will not move its output through the entire range. A 200% PB means a very small controller gain, or sensitivity to errors.

The transfer function of a proportional controller is

$$\frac{M(s)}{E(s)} = K_C \tag{3-2.5}$$

To summarize briefly, proportional controllers are the simplest controllers, with the advantage of only one tuning parameter, K_C or PB. The disadvantage of these controllers is operation with an offset in the controlled variable. In some processes,

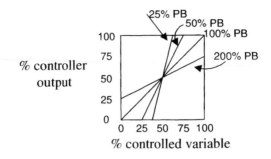

	Controller output		
	0%	50%	100%
PB = 100%	100°C	200°C	300°C
PB = 50%	150°C	200°C	250°C
PB = 25%	175°C	200°C	225°C
PB = 200%		200°C	

Figure 3-2.3   Definition of proportional band.

such as the level in a surge tank, the cruise control in a car, or a thermostat in a house, this may not be of any major consequence. For processes in which the process variable can be controlled within a band from set point, proportional controllers are sufficient. However, when the process variable must be controlled at the set point, not away from it, proportional controllers do not provide the required control.

3-2.2 Proportional–Integral Controller

Most processes cannot be controlled with an offset; that is, they must be controlled at the set point. In these instances an extra amount of "intelligence" must be added to the proportional controller to remove the offset. This new intelligence, or new mode of control, is the integral, or reset, action; consequently, the controller becomes a proportional–integral (PI) controller. The describing equation is

$$m(t) = \overline{m} + K_C e(t) + \frac{K_C}{\tau_I} \int e(t)\,dt \qquad (3\text{-}2.6)$$

where τ_I is the integral (or reset) time with units of time. Most often, the time unit used is minutes; less often, seconds are used. The unit used depends on the manufacturer. Therefore, the PI controller has two parameters, K_C and τ_I, both of which must be adjusted (tuned) to obtain satisfactory control.

To understand the physical significance of the reset time, consider the hypothetical example shown in Fig. 3-2.4. At some time $t = 0$, a constant error of 1% in magnitude is introduced in the controller. At this moment the PI controller solves the following equation:

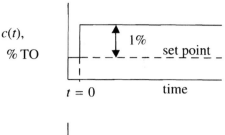

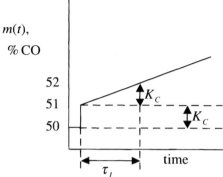

Figure 3-2.4 Response of a PI controller (direct acting) to a step change in error.

$$m(t) = 50\% + K_C e(t) + \frac{K_C}{\tau_I} \int_0^t e(t)\,dt$$

$$= 50\% + K_C(1) + \frac{K_C}{\tau_I} \int_0^t (1)\,dt$$

$$= 50\% + K_C + \frac{K_C}{\tau_I} t$$

When the error is introduced at $t = 0$, the controller output changes immediately by an amount equal to K_C; this is the response due to the proportional mode.

$$m(t = 0) = 50\% + K_C + \frac{K_C}{\tau_I}(0) = 50\% + K_C$$

As time increases the output also increases in a ramp fashion as expressed by the equation and shown in the figure. Note that when $t = \tau_I$ the controller output becomes

$$m(t = \tau_I) = 50\% + K_C + K_C$$

Thus, in an amount of time equal to τ_I, the integral mode repeats the action taken by the proportional mode. The smaller the value of τ_I, the faster the controller integrates to repeat the proportional action. Realize that the smaller the value of τ_I, the larger the term in front of the integral, K_C/τ_I, and consequently, the faster the integral term moves the controller output.

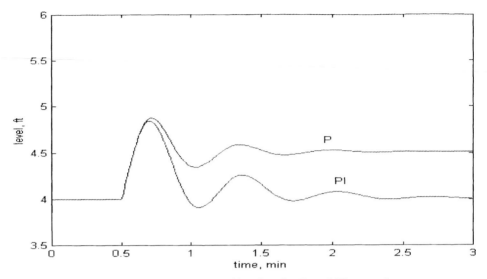

Figure 3-2.5   Response of level under P and PI control.

To explain why the PI controller removes the offset, consider the level control system used previously to explain the offset required by a P controller. Figure 3-2.5 shows the response of the level under P and PI controllers to a change in inlet flow from 150 gpm to 170 gpm. The response with a P controller shows the offset, while the response with a PI controller shows that the level returns to the set point, with no offset. Under PI control, as long as the error is present, the controller keeps changing its output (integrating the error). Once the error disappears, goes to zero, the controller does not change its output anymore (it integrates a function with a value of zero). As shown in the figure, at time = 3 min, the error disappears. The signal to the valve must still be 60%, requiring the valve to deliver 170 gpm. Let us look at the PI equation at the moment the steady state is reached:

$$m(t) = 50\% + K_C(0) + \frac{K_C}{\tau_I} \int (0)\, dt$$

$$= 50\% + 0 + 10\% = 60\%$$

The equation shows that even with a "zero" error, the integral term is not zero but, rather, 10%, which provides the required output of 60%. The fact that the error is zero does not mean that the value of the integral term is zero. It means that the integral term remains constant at the last integrated value. Integration means area under the curve, and even though the level is the same at $t = 0$ and at $t = 3$ min, the value of the integral is different (different areas under the curve) at these two times. The value of the integral term times K_C/τ_I is equal to 10%. Once the level returns to the set point, the error disappears and the integral term remains constant. *Integration is the mode that removes the offset.*

Some manufacturers do not use the reset time for their tuning parameter. They use the reciprocal of reset time, which we shall refer to as reset rate, τ_I^R; that is,

$$\tau_I^R = \frac{1}{\tau_I} \tag{3-2.7}$$

The unit of τ_I^R is therefore 1/time or simply (time)$^{-1}$. Note that when τ_I is used and faster integration is desired, a smaller value must be used in the controller. However, when τ_I^R is used, a larger value must be used. Therefore, before tuning the reset term, the user must know whether the controller uses reset time (time) or reset rate (time)$^{-1}$; τ_I and τ_I^R *are the reciprocal of one another, and consequently, their effects are opposite.*

As we learned in Section 3.2.1, two terms are used for the proportional mode (K_C and PB), and now we have just learned that there are two terms for the integral mode (τ_I and τ_I^R). This can be confusing, and therefore it is important to keep the differences in mind when tuning a controller. Equation (3-2.6), together with the following equations, show the four possible combinations of tuning parameters; we refer to Eq. (3-2.6) as the *classical controller equation.*

$$m(t) = \overline{m} + \frac{100}{PB} e(t) + \frac{100}{PB} \tau_I \int e(t)\, dt \tag{3-2.8}$$

$$m(t) = \overline{m} + \frac{100}{PB} e(t) + \frac{100}{PB} \tau_I^R \int e(t)\, dt \tag{3-2.9}$$

$$m(t) = \overline{m} + K_C e(t) + K_C \tau_I^R \int e(t)\, dt \tag{3-2.10}$$

The transfer function for the classical PI controller is

$$G_C(s) = \frac{M(s)}{E(s)} = K_C \left(1 + \frac{1}{\tau_I s} \right) = K_C \frac{\tau_I s + 1}{\tau_I s} \tag{3-2.11}$$

To summarize, proportional–integral controllers have two tuning parameters: the gain or proportional band, and the reset time or reset rate. *The advantage is that the integration removes the offset.* About 85% of all controllers in use are of this type. The disadvantage of the PI controller is related to the stability of the control loop. Remembering that the ultimate gain, K_{CU}, is considered the limit of stability (maximum value of K_C before the system goes unstable), theory predicts, and practice confirms, that for a PI controller the K_{CU} is less than for a proportional controller. That is,

$$K_{CU}|_P > K_{CU}|_{PI}$$

The addition of integration adds some amount of instability to the system; this is presented in more detail in Chapter 5. Therefore, to counteract this effect, the controller must be tuned somewhat less aggressively (smaller K_C). The formulas we use to tune controllers will take care of this.

3-2.3 Proportional–Integral–Derivative Controller

Sometimes another mode of control is added to the PI controller. This new mode of control is the *derivative action*, also called the *rate action*, or *pre-act*. Its purpose is to anticipate where the process is heading by looking at the time rate of change of the error, its derivative. The describing equation is

$$m(t) = \overline{m} + K_C e(t) + \frac{K_C}{\tau_I} \int e(t)\, dt + K_C \tau_D \frac{de(t)}{dt} \tag{3-2.12}$$

where τ_D = derivative (or rate) time. The time unit used is generally minutes; however, some manufacturers use seconds.

The proportional–integral–derivative (PID) controller has three terms, K_C or PB, τ_I or τ_I^R, and τ_D, which must be adjusted (tuned) to obtain satisfactory control. The derivative action gives the controller the capability to anticipate where the process is heading, that is, to look ahead by calculating the derivative of the error. The amount of anticipation is decided by the value of the tuning parameter, τ_D.

Let us consider the heat exchanger shown in Fig. 3-1.1 and use it to clarify what is meant by "anticipation." Assume that the inlet process temperature decreases by some amount and the outlet temperature starts to decrease correspondingly, as shown in Fig. 3-2.6. At time t_a the amount of the error is positive and small. Consequently, the amount of control correction provided by the proportional and integral modes is small. However, the derivative of this error, that is, the slope of the error curve, is large and positive, making the control correction provided by the derivative mode large. By looking at the derivative of the error, the controller knows that the controlled variable is heading away from the set point rather fast, and consequently, it uses this fact to help in controlling. At time t_b the error is still positive and larger than before. The amount of control correction provided by the proportional and integral modes is also larger than before and still adding to the output of the controller to open the steam valve further. However, the derivative of the error at this time is negative, signifying that the error is decreasing; the controlled variable has started to come back to the set point. Using this fact, the derivative mode starts to subtract from the other two modes since it recognizes that the error is decreasing. This controller results in reduced overshoot and oscillations around the set point.

PID controllers are recommended for use in slow processes (processes with long time constants), such as temperature loops, which are usually free of noise. Fast processes (processes with short time constants) are easily susceptible to process noise. Typical of these fast processes are flow loops and liquid pressure loops. For these processes with noise, the use of derivative action will amplify the noise and therefore should not be used for these processes.

The transfer function of a PID controller is given by

$$G_C(s) = \frac{M(s)}{E(s)} = K_C\left(1 + \frac{1}{\tau_I s} + \tau_D s\right) \tag{3-2.13}$$

To summarize, PID controllers have three tuning parameters: the gain or proportional band, the reset time or reset rate, and the rate time. PID controllers should

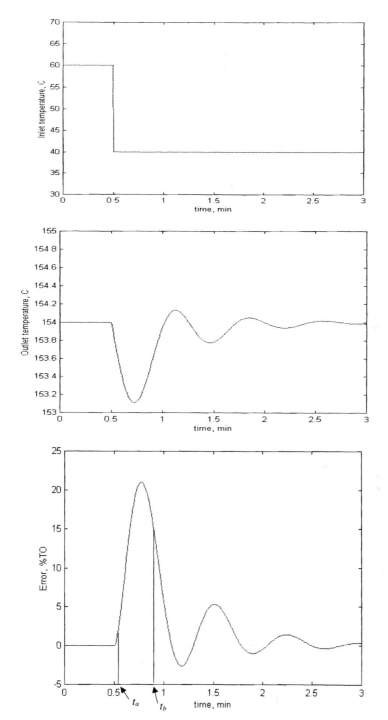

Figure 3-2.6   Response of heat exchanger temperature to a disturbance.

not be use in processes with noise. An advantage of the derivative mode is that it provides anticipation. Another advantage is related to the stability of the system. Theory predicts, and practice confirms, that the ultimate gain with a PID controller is larger than that of a PI controller. That is,

$$K_{CU}|_{PID} > K_{CU}|_{PI}$$

The derivative terms add some amount of stability to the system; this is presented in more detail in Chapter 5. Therefore, the controller can be tuned more aggressively now. The formulas we'll use to tune controllers will take care of this.

3-2.4 Proportional–Derivative Controller

The proportional–derivative (PD) controller is used in processes where a proportional controller can be used, where steady-state offset is acceptable but some amount of anticipation is desired, and no noise is present. The describing equation is

$$m(t) = \overline{m} + K_C e(t) + K_C \tau_D \frac{de(t)}{dt} \tag{3-2.14}$$

and the transfer function is

$$G_C(s) = \frac{M(s)}{E(s)} = K_C(1 + \tau_D s) \tag{3-2.15}$$

Based on our previous presentation on the effect of each tuning parameter on the stability of systems, the reader can complete the following:

$$K_{CU}|_{PD} \quad ? \quad K_{CU}|_{P}$$

3-3 RESET WINDUP

The problem of reset windup is an important and realistic one in process control. It may occur whenever a controller contains integration. The heat exchanger control loop shown in Fig. 3-1.1 is again used at this time to explain the reset windup problem.

Suppose that the process inlet temperature drops by an unusually large amount; this disturbance drops the outlet temperature. The controller (PI or PID) in turn asks the steam valve to open. Because the valve is fail-closed, the signal from the controller increases until, because of the reset action, the outlet temperature equals the desired set point. But suppose that in the effort of restoring the controlled variable to the set point, the controller integrates up to 100% because the drop in inlet temperature is too large. At this point the steam valve is wide open and therefore the control loop cannot do any more. Essentially, the process is out of control; this is shown in Fig. 3-3.1. The figure consists of four graphs: the inlet temperature, the outlet temperature, the valve position, and the controller's output. The figure shows

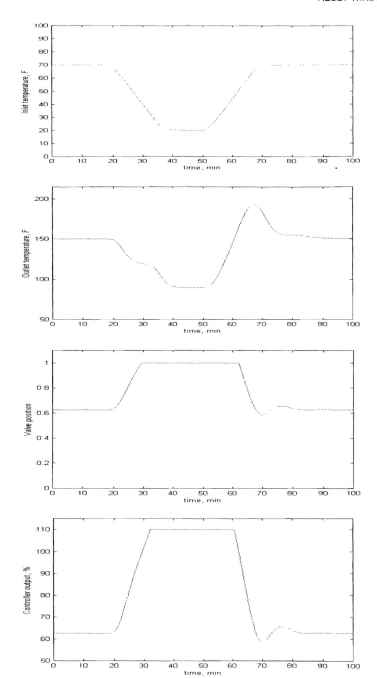

Figure 3-3.1   Heat exchanger control, reset windup.

that when the valve is fully open, the outlet temperature is not at set point. Since there is still an error, the controller will try to correct for it by further increasing (integrating the error) its output even though the valve will not open more after 100%. The output of the controller can in fact integrate above 100%. Some controllers can integrate between −15 and 115%, others between −7 and 107%, and still others between −5 and 105%. Analog controllers can also integrate outside their limits of 3 to 15 psig or 4 to 20 mA. Let us suppose that the controller being used can integrate up to 110%; at this point the controller cannot increase its output anymore; its output has become saturated. This state is also shown in Fig. 3-3.1. This saturation is due to the reset action of the controller and is referred to as *reset windup*.

Suppose now that the inlet temperature goes back up; the outlet process temperature will in turn start to increase, as also shown in Fig. 3-3.1. The outlet temperature reaches and passes the set point and the valve remains wide open when, in fact, it should be closing. The reason the valve is not closing is because the controller must now integrate from 110% down to 100% before it starts to close. Figure 3-3.2 shows an expanded view of how the controller's output starts to decrease from 110% and reaches 100% before the valve actually starts to close. The

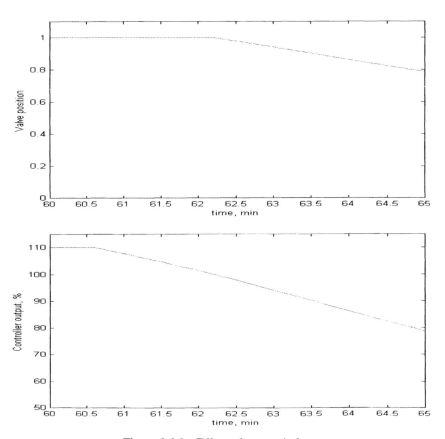

Figure 3-3.2   Effect of reset windup.

figure shows that it takes about 1.5 min for the controller to integrate down to 100%; all this time the valve is wide open. By the time the valve starts to close, the outlet temperature has overshot the set point by a significant amount, about 30°F in this case.

As mentioned earlier, this problem of reset windup may occur whenever integration is present in the controller. It can be avoided if the integration is limited to 100% (or 0%). Note that the prevention of reset windup requires us to limit the integration, not to limit the controller output when its value reaches 100% or 0%. While the output does not go beyond the limits, the controller may still be internally wound up, because it is the integral mode that winds up. Reset windup protection is an option that must be bought in analog controllers; however, it is a standard feature in DCS controller.

Reset windup occurs any time a controller is not in charge, such as when a manual bypass valve is open or when there is insufficient manipulated variable power. It also typically occurs in batch processes, in cascade control, and when a final control element is driven by more than one controller, as in override control schemes. Cascade control is presented in Chapter 4, and override control is presented in Chapter 5.

3-4 TUNING FEEDBACK CONTROLLERS

Probably 80 to 90% of feedback controllers are tuned by instrument technicians or control engineers based on their previous experience. For the 10 to 20% of cases where no previous experience exists, or for personnel without previous experience, there exist several organized techniques to obtain a "good ballpark figure" close to the "optimum" settings.

To use these organized procedures, we must first obtain the characteristics of the process. Then, using these characteristics, the tunings are calculated using simple formulas; Fig. 3-4.1 depicts this concept. There are two ways to obtain the process characteristics, and consequently, we divide the tuning procedures into two types: *on-line* and *off-line*.

3-4.1 Online Tuning: Ziegler–Nichols Technique [1]

The Ziegler–Nichols technique is the oldest method for online tuning. It gives approximate values of the tuning parameters K_C, τ_I, and τ_D to obtain approximately a one fourth ($\frac{1}{4}$) decay ratio response. The procedure is as follows:

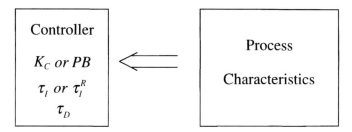

Figure 3-4.1   Tuning concept.

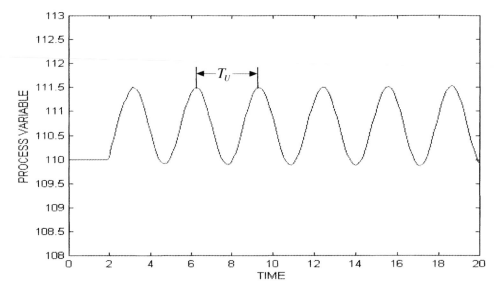

Figure 3-4.2   Testing for Ziegler–Nichols method.

1. With the controller online (in automatic), remove all the reset (τ_I = maximum or τ_I^R = minimum) and derivative ($\tau_D = 0$) modes. Start with a small K_C value.
2. Make a small set point or load change and observe the response.
3. If the response is not continuously oscillatory, increase K_C, or decrease PB, and repeat step 2.
4. Repeat step 3 until a continuous oscillatory response is obtained.

The gain that gives these continuous oscillations is the *ultimate gain*, K_{CU}. The period of the oscillations is called the *ultimate period*, T_U; this is shown in Fig. 3-4.2. The ultimate gain and the ultimate period are the characteristics of the process being tuned. The following formulas are then applied:

- For a P controller: $K_C = 0.5K_{CU}$
- For a PI controller: $K_C = 0.45K_{CU}$, $\tau_I = T_U/1.2$
- For a PID controller: $K_C = 0.65K_{CU}$, $\tau_I = T_U/2$, $\tau_D = T_U/8$

Figure 3-4.3 shows the response of a process with a PI controller tuned by the Ziegler–Nichols method. The figure also shows the meaning of a $\frac{1}{4}$ decay ratio response.

3-4.2 Offline Tuning

The data required for the offline tuning techniques are obtained from the step testing method presented in Chapter 2, that is, from K, τ, and t_o. Remember that K must be in %TO/%CO, and τ and t_o in time units consistent with those used in the controller to be tuned. These three terms describe the characteristics of the process. Once the data are obtained, any of the methods described below can be applied.

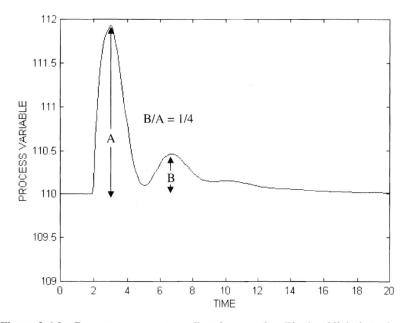

Figure 3-4.3 Process response to a disturbance using Ziegler–Nichols tunings.

Ziegler–Nichols Method [2]. The Ziegler–Nichols settings can also be obtained from the following formulas:

- For a P controller: $K_C = (1/K)(t_o/\tau)^{-1}$
- For a PI controller: $K_C = (0.9/K)(t_o/\tau)^{-1}$, $\tau_I = 3.33t_o$
- For a PID controller: $K_C = (1.2/K)(t_o/\tau)^{-1}$, $\tau_I = 2.0t_o$, $\tau_D = 0.5t_o$

In Section 2-3 we presented the meaning of dead time. We mentioned that the dead time has an adverse effect on the controllability of processes. Furthermore, the larger the dead time with respect to the time constant, the less aggressive the controller will have to be tuned. The Ziegler–Nichols tuning formulas clearly show this dependence on dead time. The formulas show that the larger the t_o/τ ratio, the smaller the K_C. Chapter 5 presents further proof of the adverse effects of dead time.

The Ziegler–Nichols method was developed for $t_o/\tau < 1.0$. For ratios greater than 1.0, the tunings obtained by this method become very conservative.

Controller Synthesis Method [2]. The controller synthesis method (CSM) was introduced by Martin, Corripio, and C. L. Smith [3]. Several years later *internal model control* (IMC) [4] was presented and the tunings from this method agree with those from the CSM. Some people also refer to the CSM as the *lambda tuning method.*

- For a P controller: $K_C = \tau/K(\lambda + t_o)$
- For a PI controller: $K_C = \tau/K(\lambda + t_o)$, $\tau_I = \tau$
- For a PID controller: $K_C = \tau/K(\lambda + t_o)$, $\tau_I = \tau$, $\tau_D = t_o/2$

Looking at the formulas, it is clear that for each controller it comes down to only one tuning parameter, λ. As the formulas show, the smaller the λ value, the more aggressive (the larger the K_C) the controller becomes. We recommend the following values of λ as a first guess:

- For a P controller: $\lambda = 0$
- For a PI controller: $\lambda = t_o$
- For a PID controller: $\lambda = 0.2t_o$

The response obtained by this method tend to give a more overdamped (less oscillatory) response than the Ziegler–Nichols, depending on the value of λ used. Figure 3-4.4 shows the response of the same process as in Fig. 3-4.3, but this time with a PI controller tuned by the CSM method, with the λ suggested. The CSM method is not limited by the value of t_o/τ as are the Ziegler–Nichols tunings.

Other Tunings. In this section we discuss the tuning of flow loops and level loops. Both loops are quite common, and present characteristics that make it difficult to tune them with the methods presented thus far.

Flow Loops. Flow loops are the most common loops in the process industries. Their dynamic response is rather fast. Consider the loop shown in Fig. 3-4.5. Assume that the controller is in manual and a step change in controller output is induced. The response of the flow is almost instantaneous; the only dynamic element is the control valve. The two-point method of Chapter 2, used to obtain a first-order-plus-dead-time approximation of the response, shows that the dead-time term is very close to zero, $t_o \approx 0$ min. In every tuning equation for controller gain, the dead time appears

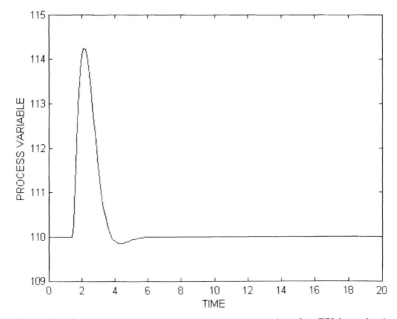

Figure 3-4.4 Process response to a disturbance using the CSM method.

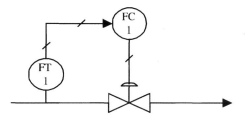

Figure 3-4.5 Flow loop.

in the denominator of the equation. Thus the results would show a need for an infinite controller gain. Analysis of these types of fast processes [2] indicates that the controller needed is an integral only. Because pure integral controllers were not available when only analog instrumentation was available, a PI controller was used with very small proportional action and a large integral action. Today, this practice is still followed. The following is offered as a rule of thumb for flow loops:

- Conservative tuning: $K_C = 0.1$, $\tau_I = 0.1$ min
- Aggressive tuning: $K_C = 0.2$, $\tau_I = 0.05$ min

Note what these tunings offer. Consider the equation for a PI controller, Eq. (3-2.6):

$$m(t) = \overline{m} + K_C e(t) + \frac{K_C}{\tau_I} \int e(t)\, dt$$

The conservative tunings provide a proportional action, $K_C = 0.1$, and an integral action, $K_C/\tau_I = 0.1/0.1 = 1.0$, or 10 times more integral action than proportional action. The aggressive tunings provide 20 times more integral action than proportional action. Thus the PI controller is used to approximate an integral controller.

In Chapter 4 we discuss cascade control. Flow loops are commonly used as "slave loops" in cascade control. In these cases, flow controllers with a gain of 0.9 give better overall response. Remember this when you read Chapter 4.

Level Loops. Level loops present two interesting characteristics. The first characteristic is that as presented in Chapter 2, very often levels are integrating processes. In this case it is impossible to obtain a response to approximate it with a first-order-plus-dead-time model. That is, it is impossible to obtain K, τ, and t_o, and therefore we cannot use any tuning equation presented thus far. Those levels processes that are not integrating processes but rather, self-regulating processes can be approximated by a first-order-plus-dead-time model, as shown in this chapter.

The second characteristic of level loops is that there are two possible *control objectives*. To explain these control objectives, consider Fig. 3-4.6. If the input flow varies as shown in the figure, to control the level tightly at set point the output flow must also vary, as shown. We referred to this as *tight level control*. However, the

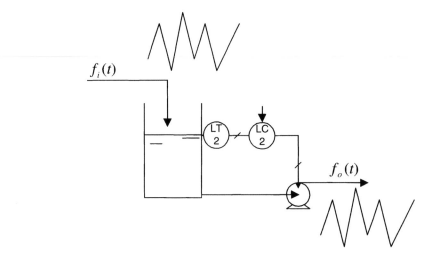

Figure 3-4.6 Level loop.

changes in output flow will act as disturbances to the downstream process unit. If this unit is a reactor, separation column, filter, and so on, the disturbance may have a major effect on its performance. Often, it is desired to smooth the flow feeding the downstream unit. To accomplish this objective, the level in the tank must be allowed to "float" between a high and a low level. Thus, the objective is not to control the level tightly but rather, to smooth the output flow with some consideration of the level. We referred to this objective as *average level control*. Let us look at how to tune the level controller for each objective.

TIGHT LEVEL CONTROL. If the level process happens to be self-regulated, that is, if it is possible to obtain K, τ, and t_o, the tuning techniques already presented in this chapter can be used. If the level process is integrating, the following equation [2] for a proportional controller is proposed:

$$K_C = \frac{A}{4\tau_V K_V K_T} \tag{3-4.1}$$

where, A is the cross-sectional area of tank (length^2), τ_V the time constant of the valve (time), K_V the valve's gain [length^3/(time · %CO)], and K_T the transmitter's gain (%TO/length). The valve's gain can be approximated by

$$K_V = \frac{\text{maximum volumetric flow provided by valve}}{100\%CO}$$

The transmitter's gain can be calculated by

$$K_T = \frac{100\%TO}{\text{transmitter's span}}$$

The time constant of the valve depends on several things, such as the size of the actuator, whether a positioner is used or not, and so on. Anywhere between 3 and 10 seconds (0.05 to 0.17 minutes) could be used. For a more in-depth development and discussion, see Ref. 2, pp. 334 and 335.

AVERAGE LEVEL CONTROL. To review what we had previously said, the objective of average level control is to smooth the output flow from the tank. To accomplish this objective, the level in the tank must be allowed to "float" between a high and a low level. Obviously, the larger the difference between the high and low levels, the more "capacitance" is provided, and the more smoothing of the flow is obtained.

There are two ways to tune a proportional controller for average level control. The first way is also discussed in Ref. 2 and says: *The ideal averaging level controller is a proportional controller with the set point at 50% TO, the output bias at 50% CO, and the gain set at 1% CO/% TO.* The tuning obtained in this case results in that the level in the tank will vary the full span of the transmitter as the valve goes from wide open to completely closed. Thus the full capacitance provided by the transmitter is used.

To explain the second way to tune the controller, consider Fig. 3-4.7. The figure shows two deviations, D1 and D2, not present in Fig. 3-4.6. D1 indicates the *expected flow deviation from the average flow*. D2 indicates the *allowed level deviation from set point*. With this information we can now write the tuning equation:

$$K_C = (0.75)\frac{\dfrac{\text{expected flow deviation from the average input flow (D1)}}{\text{maximum flow given by final control element } (f_{o.\,\text{max}})}}{\dfrac{\text{allowed level deviation from set point (D2)}}{\text{span of level transmitter}}} \qquad (3\text{-}4.2)$$

The equation is composed of two ratios, and both ratios must be dimensionless. This equation allows you to (1) use less than the span of the transmitter if it is necessary for some reason, and (2) take into consideration the variations in input flow. For best results, the level should be allowed to vary as much as possible and

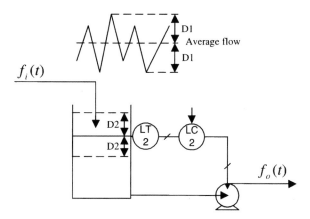

Figure 3-4.7 Level loop.

D2 made as large as possible; this is a decision for the engineer. D1 depends on the process. The final control element shown in Fig. 3-4.3 is a pump; valves are also common. The equation was developed so that once the engineer decides on D2, this limit is not violated, while providing smoothing of the output flow.

3-5 SUMMARY

In this chapter we have seen that the purpose of controllers is *to make decisions on how to use the manipulated variable to maintain the controlled variable at set point.* We have discussed the significance of the action of the controller, reverse or direct, and how to select the appropriate one. The carious types of controllers were also studied, stressing the significance of the tuning parameters, gain K_C or proportional band PB, reset time τ_I or reset rate τ_I^R, and derivative or rate time τ_D. The subject of reset windup was presented and its significance discussed. Finally, various tuning techniques were presented and discussed.

REFERENCES

1. J. G. Ziegler and N. B. Nichols, Optimum setting for automatic controllers, *Transactions ASME*, 64:759, November 1942.
2. C. A. Smith and A. B. Corripio, *Principles and Practice of Automatic Process Control*, Wiley, New York, 1997.
3. J. Martin, Jr., A. B. Corripio, and C. L. Smith, How to select controller modes and tuning parameters from simple process models, *ISA Transactions*, 15(4):314–319, 1976.
4. D. E. Rivera, M. Morari, and S. Skogestad, Internal model control: 4. PID controller design, *I&EC Process Design and Development*, 25:252, 1986.

PROBLEMS

3-1. For the process of Problem 2-1, decide on the action of the controller and tune a PID controller.

3-2. For the process of Problem 2-2, decide on the action of the controller and tune a PI controller.

CHAPTER 4

CASCADE CONTROL

Feedback control is the simplest strategy of automatic process control that compensates for process upsets. However, the disadvantage of feedback control is that it reacts only after the process has been upset. That is, if a disturbance enters the process, it has to propagate through the process, make the controlled variable deviate from the set point, and it is then that feedback takes corrective action. Thus a deviation in the controlled variable is needed to initiate corrective action. Even with this disadvantage, probably 80% of all control strategies used in industrial practice are simple feedback control. In these cases the control performance provided by feedback is satisfactory for safety, product quality, and production rate.

As process requirements tighten, or in processes with slow dynamics, or in processes with too many or frequently occurring upsets, the control performance provided by feedback control may become unacceptable. Thus it is necessary to use other strategies to provide the performance required. These additional strategies are the subject of this and some of the subsequent chapters. The strategies presented complement feedback control; they do not replace it. The reader must remember that it is always necessary to provide some feedback from the controlled variable.

Cascade control is a strategy that in some applications improves significantly the performance provided by feedback control. This strategy is well known and has been used for a long time. The fundamentals and benefits of cascade control are explained in detail in this chapter.

4-1 PROCESS EXAMPLE

Consider the furnace/preheater and reactor process shown in Fig. 4-1.1. In this process a well-known reaction, A → B, occurs in the reactor. Reactant A is usually available at a low temperature; therefore, it must be heated before being fed to the

61

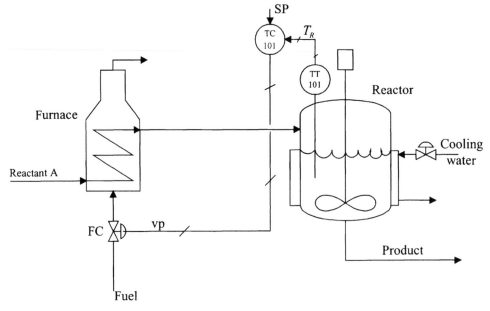

Figure 4-1.1 Feedback control of reactor.

reactor. The reaction is exothermic, and to remove the heat of reaction, a cooling jacket surrounds the reactor.

The important controlled variable is the temperature in the reactor, T_R. The original control strategy called for controlling this temperature by manipulating the flow of coolant to the jacket. The inlet reactant temperature to the reactor was controlled by manipulating the fuel valve. It was noted during the startup of this process that the cooling jacket could not provide the cooling capacity required. Thus it was decided to open the cooling valve completely and control the reactor temperature by manipulating the fuel to the preheater, as shown in Fig. 4-1.1. This strategy worked well enough, providing automatic control during startup.

Once the process was "lined-out," the process engineer noticed that every so often the reactor temperature would move from the set point enough to make off-spec product. After checking the feedback controller tuning to be sure that the performance obtained was the best possible, the engineer started to look for possible process disturbances. Several upsets were found around the reactor itself (cooling fluid temperature and flow variations) and others around the preheater (variations in inlet temperature of reactant A, in the heating value of fuel, in the inlet temperature of combustion air, and so on). Furthermore, the engineer noticed that every once in a while the inlet reactant temperature to the heater would vary by as much as 25°C, certainly a major upset.

It is fairly simple to realize that the effect of an upset in the preheater results first in a change of the reactant exit temperature from the preheater, T_H, and that this then affects the reactor temperature, T_R. Once the controller senses the error in T_R, it manipulates the signal to the fuel valve. However, with so many lags in the process, preheater plus reactor, it may take a considerable amount of time to bring the reactor temperature back to set point. Due to these lags, the simple feedback control shown in the figure will result in cycling, and in general, sluggish control.

A superior control strategy can be designed by making use of the fact that the upsets in the preheater first affect T_H. Thus it is logical to start manipulating the fuel valve as soon as a variation in T_H is sensed, before T_R starts to change. That is, the idea is not to wait for an error in T_R to start changing the manipulated variable. This corrective action uses an intermediate variable, T_H in this case, to reduce the effect of some dynamics in the process. This is the idea behind cascade control, and it is shown in Fig. 4-1.2.

This strategy consists of two sensors, two transmitters, two controllers, and one control valve. One sensor measures the intermediate, or secondary, variable T_H in this case, and the other sensor measures the primary controlled variable, T_R. Thus this strategy results in two control loops, one loop controlling T_R and the other loop controlling T_H. To repeat, the preheater exit temperature is used only as an intermediate variable to improve control of the reactor temperature, which is the important controlled variable.

The strategy works as follows: Controller TC101 looks at the reactor temperature and decides how to manipulate the preheater outlet temperature to satisfy its set point. This decision is passed on to TC102 in the form of a set point. TC102, in turn, manipulates the signal to the fuel valve to maintain T_H at the set point given by TC101. If one of the upsets mentioned earlier enters the preheater, T_H deviates from the set point and TC102 takes corrective action right away, before T_R changes. Thus the dynamic elements of the process have been separated to compensate for upsets in the heater before they affect the primary controlled variable.

In general, the controller that keeps the primary variable at set point is referred to as the *master controller, outer controller,* or *primary controller.* The controller

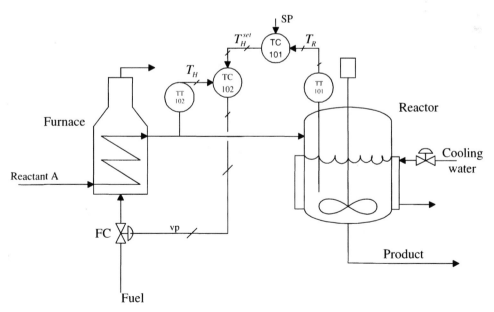

Figure 4-1.2   Cascade control of reactor.

used to maintain the secondary variable at the set point provided by the master controller is usually referred to as the *slave controller*, *inner controller*, or *secondary controller*. The terminology *primary/secondary* is commonly preferred because for systems with more than two cascaded loops, it extends naturally.

Note that the secondary controller receives a signal from the primary controller and this signal is used as the set point. To "listen" to this signal, the controller must be set in what is called *remote set point* or *cascade*. If one desires to set the set point manually, the controller must then be set in *local set point* or *auto*.

Figure 4-1.3 shows the response of the process to a –25°C change in inlet reactant temperature, under simple feedback control, and under cascade control. The improvement is very significant and in all probability in this case pays for the added expenses in no time.

The following must be stressed: *In designing cascade control strategies, the most important consideration is that the secondary variable must respond faster to changes in the disturbance, and in the manipulated variable, than the primary variable does—the faster the better*. This requirement makes sense and it is extended to any number of cascade loops. In a system with three cascaded loops, as shown in Section 4-3.2, the tertiary variable must be faster than the secondary variable, and this variable in turn must be faster than the primary variable. Note that the most inner controller is the one that sends its outputs to the valve. The outputs of all other controllers are used as set points to other controllers; for these controllers, their final control element is the set point of another controller.

As noted from this example, we are starting to develop more complex control schemes than simple feedback. It is helpful in developing these schemes, and others

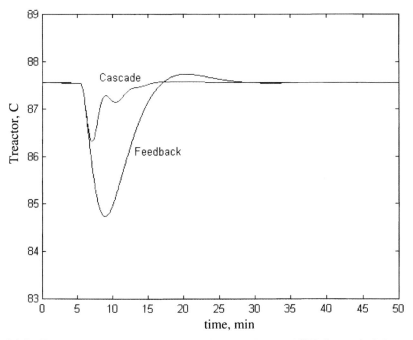

Figure 4-1.3 Response of feedback and cascade control to a –25°C change in inlet reactant temperature.

shown in the following chapters, to remember that every signal must have a physical significance. In Figs. 4-1.1 and 4-1.2 we have labeled each signal with its significance. For example, in Fig. 4-1.2 the output signal from TT101 indicates the temperature in the reactor, T_R; the output signal from TT102 indicates the outlet temperature from the heater, T_H; and the output signal from TC101 indicates the required temperature from the heater, T_H^{set}. Even though indicating the significance of the signals in control diagrams is not standard practice, we will continue to do so. This practice helps in understanding control schemes, and we recommend that the reader do the same.

4-2 IMPLEMENTATION AND TUNING OF CONTROLLERS

Two important questions remain concerning how to put the cascade strategy into full automatic operation and how to tune the controllers. The answer to both questions is the same: from inside out. That is, the inner controller is first tuned and set into remote set-point mode while the other loops are in manual. As the inner controller is set in remote set point, it is good practice to check how it performs before proceeding to the next controller. This procedure continues outwardly until all controllers are in operation. For the process shown in Fig. 4-1.2, TC102 is first tuned while TC101 is in manual. The control performance of TC102 is then checked before proceeding to TC101. This checking can usually be done very simply by varying the set point to TC102. Remember, it is desired to make TC102 as fast as possible, even if it oscillates a bit, to minimize the effect of the upsets. Once this is done, TC102 is set in remote set point, TC101 is tuned and set in automatic.

Tuning cascade control systems is more complex than simple feedback systems if for no other reason than simply because there is more than one controller to tune. However, this does not mean that it is difficult either. We first present the methods available to tune two-level cascade systems and then proceed by discussing the tuning methods available to tune three-level cascade systems.

4-2.1 Two-Level Cascade Systems

The control system shown in Fig. 4-1.2 is referred to as a *two-level cascade system*. Realize that the inner loop by itself is a simple feedback loop. Therefore, TC102 can be tuned by any of the techniques discussed in Chapter 3. As mentioned previously, the recommendation is to tune this controller as fast as possible, avoiding instability of course. The objective is to make the inner loop fast and responsive, to minimize the effect of upsets on the primary controlled variable. Tuning this system is then reduced to tuning the primary controller.

There are several way to obtain a first guess as to the tuning of the primary controller. Trial and error is often used by experienced personnel. The other methods available follow a "recipe" to obtain the first tuning values. The first method available is the Ziegler–Nichols oscillatory technique presented in Chapter 3. That is, after removing any integral or derivative action present in the primary controller, its gain is increased cautiously until the controlled variable oscillates with sustained oscillations. The controller gain that provides these oscillations is called the ultimate gain, K_{C_U}, and the period of the oscillations is the ultimate period, T_U. The Ziegler–Nichols equations presented in Chapter 3 are then used.

The second method available is the one presented by Pressler [1]. Pressler's method was developed assuming that the secondary controller is a proportional only and that the primary controller is a proportional integral; this P/PI combination is usually quite convenient. The method works well; however, it assumes that the inner loop does not contain dead time, which limits its application to cascade systems with flow or liquid pressure loops as the inner loop. For processes with dead time in the inner loop, such as the one shown in Fig. 4-1.2, the application of Pressler's method would yield an unstable response if the master controller were ever set in manual.

The third method available is to extend the offline methods presented in Chapter 3 to both primary and secondary controllers. That is, with the secondary controller in manual, a step change in its output is introduced and the response of the temperature out of the heater (secondary variable) is recorded. From the data a gain, time constant, and dead time for the secondary loop is obtained and the controller tuned by whatever method presented in Chapter 3 the engineer desires. Once this is done, the secondary controller is set in remote set point. With the primary controller in manual, a step change in its output is then introduced and the response of the reactor's temperature (primary variable) is recorded. From the data a gain, time constant, and dead time for the primary loop are obtained and the controller tuned by whatever method presented in Chapter 3 the engineer desires.

The fourth method available to tune cascade systems is the one developed by Austin [2]. The method provides a way to tune both the primary and secondary with only one step test. Tuning equations are provided for the primary controller, PI or PID, when the secondary controller is either P or PI. The method consists of generating a step change in signal to the control valve as explained in Chapter 3, and recording the response of the secondary and primary variables. The response of the secondary variable is used to calculate the gain, $K_2 = \%TT102/\%CO$, time constant τ_2, and dead time t_{o2} of the inner loop. The response of the primary variable is used to calculate the gain, $K_1 = \%TT101/\%CO$, time constant τ_1, and dead time t_{o1} of the primary loop. This information and the equations presented in Table 4-2.1 or 4-2.2 are used to obtain the tunings of the primary controller. Table 4-2.1 presents the equations to tune the primary controller when its set point is constant. However, when the set point to the primary controller is continuously changing with time, the equations provided in Table 4-2.2 are then used. Note, however, that if $\tau_2/\tau_1 > 0.38$, Table 4-2.2 should be used even if the set point to the primary controller never changes. Under this ratio condition, the equations in Table 4-2.2 provide better tunings. *The τ_2/τ_1 ratio should always be checked first.* Note that the term K_{C_2} in the tables refers to the gain of the secondary controller.

The response under cascade control shown in Fig. 4-1.3 was obtained with controller tunings calculated using Austin's method. This method provides a simple procedure to obtain near-optimum tunings for the primary controller. The fact that both controllers can be tuned from information obtained from the same test makes the method even more useful.

Figure 4-3.2a, presented in Section 4-3, shows a temperature controller cascaded to a flow controller. Cascade systems with flow controllers in the inner loop are very common and thus worthy of discussion. Following the previous presentation, after a change in the flow controller output is introduced, and the flow and temperature are recorded, the respective gains, time constants, and dead times can be obtained.

TABLE 4-2.1   Tuning Equations for Two-Level Cascade System: Disturbance Changes[a]

Secondary: P
Primary: PI   PID

$$K_{C_1} = 1.4 \left(\frac{1 + K_{C_2} K_2}{K_{C_2} K_1} \right) \left(\frac{t_{o_1}}{\tau_1} \right)^{-1.14} \left(\frac{\tau_2}{\tau_1} \right)^{0.1}$$ $$K_{C_1} = 1.4 \left(\frac{1 + K_{C_2} K_2}{K_{C_2} K_1} \right) \left(\frac{t_{o_1}}{\tau_1} \right)^{-1.14} \left(\frac{\tau_2}{\tau_1} \right)^{0.1}$$

$$\tau_{I_1} = \tau_1$$ $$\tau_{I_1} = \tau_1, \quad \tau_{D_1} = \frac{t_{o_1} - \tau_2}{2}$$

Secondary: PI
Primary: PI   PID

$$K_{C_1} = 1.25 \left(\frac{K_2}{K_1} \right) \left(\frac{t_{o_1}}{\tau_1} \right)^{-1.07} \left(\frac{\tau_2}{\tau_1} \right)^{0.1}$$ $$K_{C_1} = 1.25 \left(\frac{K_2}{K_1} \right) \left(\frac{t_{o_1}}{\tau_1} \right)^{-1.07} \left(\frac{\tau_2}{\tau_1} \right)^{0.1}$$

$$\tau_{I_1} = \tau_1$$ $$\tau_{I_1} = \tau_1, \quad \tau_{D_1} = \frac{t_{o_1} - \tau_2}{2}$$

Range: $0.02 \le \dfrac{\tau_2}{\tau_1} \le 0.38$ Range: $0.02 \le \dfrac{\tau_2}{\tau_1} \le 0.38$

$$\frac{t_{o_2}}{t_{o_1}} \le 1.0$$ $$t_{o_2} \le t_{o_1}$$

$$\frac{t_{o_1} - \tau_2}{2} \ge 0.08$$

[a]If $\tau_2/\tau_1 > 0.38$, use Table 4-2.2.

Since flow loops are quite fast, the time constant will be on the order of seconds and the dead time very close to zero, $t_{o_2} \approx 0$ min. As presented in Section 3-4.2, flow controllers are usually tuned with low gain, $K_C \approx 0.2$, and short reset time, $\tau_I \approx 0.05$ min. However, in the process shown in Fig. 4-3.2*a*, the flow controller is the inner controller in a cascade system, and because it is desired to have a fast-responding inner loop, the recommendation in this case is to increase the controller gain close to 1, $K_C \approx 1.0$; to maintain stability, the reset time may also have to be increased [3]. Once the flow controller has been tuned to provide fast and stable response, the temperature controller can be tuned following Austin's guidelines. It is important to realize that t_{o_2} is not a factor in the equations and therefore will not have an effect in the tuning of the master controller.

Another method to tune the cascade loop of a temperature controller cascaded to a flow controller, as in a heat exchanger, is to reduce the two-level cascade system to a simple feedback loop by realizing that the flow loop is very fast and thus just considering it as part of the valve, and therefore as part of the process. This is done by first tuning the flow controller as explained previously and setting it in remote set point. Once this is done, the flow controller is receiving its set point from the temperature controller. Then introduce a step change from the temperature controller and record the temperature. From the recording calculate the gain, time con-

TABLE 4-2.2   Tuning Equations for Two-Level Cascade System: Set-Point Changes

Secondary: P

Primary:	PI	PID
	$K_{C_1} = 0.84 \left(\dfrac{1 + K_{C_2} K_2}{K_{C_2} K_1} \right) \left(\dfrac{t_{o1}}{\tau_1} \right)^{-1.14} \left(\dfrac{\tau_2}{\tau_1} \right)^{0.1}$	$K_{C_1} = 1.17 \left(\dfrac{1 + K_{C_2} K_2}{K_{C_2} K_1} \right) \left(\dfrac{t_{o1}}{\tau_1} \right)^{-1.14} \left(\dfrac{\tau_2}{\tau_1} \right)^{0.1}$
	$\tau_{I_1} = \tau_1$	$\tau_{I_1} = \tau_1, \quad \tau_{D_1} = \dfrac{t_{o1} - \tau_2}{2}$

Secondary: PI

Primary:	PI	PID
	$K_{C_1} = 0.75 \left(\dfrac{K_2}{K_1} \right) \left(\dfrac{t_{o1}}{\tau_1} \right)^{-1.07} \left(\dfrac{\tau_2}{\tau_1} \right)^{0.1}$	$K_{C_1} = 1.04 \left(\dfrac{K_2}{K_1} \right) \left(\dfrac{t_{o1}}{\tau_1} \right)^{-1.07} \left(\dfrac{\tau_2}{\tau_1} \right)^{0.1}$
	$\tau_{I_1} = \tau_1$	$\tau_{I_1} = \tau_1, \quad \tau_{D_1} = \dfrac{t_{o1} - \tau_2}{2}$
	Range: $0.02 \le \dfrac{\tau_2}{\tau_1} \le 0.65$	Range: $0.02 \le \dfrac{\tau_2}{\tau_1} \le 0.35$
	$\dfrac{t_{o2}}{t_{o1}} \le 1.0$	$t_{o2} \le t_{o1}$
		$\dfrac{t_{o1} - \tau_2}{2} \ge 0.08$

stant, and dead time. With this information, tune the temperature controller by any of the methods presented in Chapter 3.

4-2.2 Three-Level Cascade Systems

Controller TC102 in the cascade system shown in Fig. 4-1.2 manipulates the valve position to maintain the preheater outlet temperature at set point. The controller manipulates the valve position, not the fuel flow. The fuel flow depends on the valve position and on the pressure drop across the valve. A change in this pressure drop, a common upset, results in a change in fuel flow. The control system, as is, will react to this upset once the outlet preheater temperature deviates from the set point. If it is important to minimize the effect of this upset, tighter control can be obtained by adding one extra level of cascade, as shown in Fig. 4-2.1. The fuel flow is then manipulated by TC102, and a change in flow, due to pressure drop changes, would then be corrected immediately by FC103. The effect of the upset on the outlet preheater temperature would be minimal.

In this new three-level cascade system, the most inner loop, the flow loop, is the fastest. Thus the necessary requirement of decreasing the loop speed from "inside out" is maintained. To tune this three-level cascade system, note that controllers FC103 and TC102 constitute a two-level cascade subsystem in which the inner con-

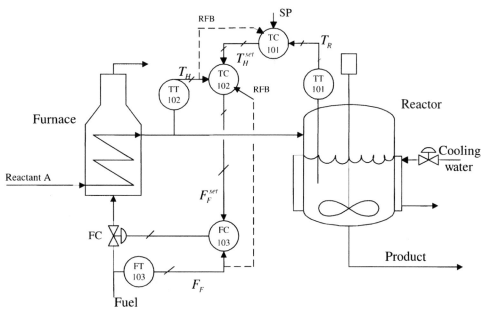

Figure 4-2.1 Three-level cascade system.

troller is very fast. Furthermore, this is exactly the case just described at the end of Section 4-2.1. Following this discussion, the flow controller is first tuned and set in remote set point. Thus, tuning this three-level cascade system reduces to tuning a two-level cascade system. Austin's method is very easily applied. With TC101 and TC102 in manual and FC103 in remote set point, introduce a step change in the signal from TC102 to FC103, and record the furnace and reactor temperatures responses. From the furnace temperature response, obtain the gain K_2 (%TT102/%CO), the time constant τ_2, and the dead time t_{o2}. Using the reactor temperature response, obtain the gain K_1 (%TT101/%CO), the time constant τ_1, and the dead time t_{o1}. With K_2, τ_2, and t_{o2}, tune the secondary controller using the equations presented in Chapter 3. Then use Table 4-2.1 or 4-2.2 to tune the primary controller.

4-3 OTHER PROCESS EXAMPLES

Consider the heat exchanger control system shown in Fig. 4-3.1, in which the outlet process fluid temperature is controlled by manipulating the steam valve position. In Section 4-2 we stated that the flow through any valve depends on the valve position *and* on the pressure drop across the valve. If a pressure surge in the steam pipe occurs, the steam flow will change. The temperature control loop shown can compensate for this disturbance only after the process temperature has deviated away from the set point.

Two cascade schemes that improve this temperature control, when steam pressure surges are important disturbances, are shown in Fig. 4-3.2. Figure 4-3.2a shows a cascade scheme in which a flow loop has been added; the temperature controller

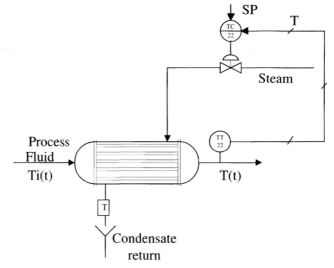

Figure 4-3.1   Temperature control.

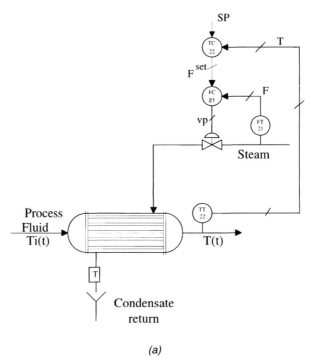

(a)

Figure 4-3.2   Cascade control schemes in heat exchanger temperature control.

resets the flow controller set point. Any flow changes are now compensated by the flow loop. The cascade scheme shown in Fig. 4-3.2*b* accomplishes the same control, but now the secondary variable is the steam pressure in the exchanger shell side. Any change in steam flow quite rapidly affects the shell-side pressure. Any pressure change is then compensated by the pressure loop. This pressure loop also compensates for disturbances in the heat content (superheat and latent heat) of the steam, since the pressure in the shell side is related to the condensing temperature and thus to the heat transfer rate in the exchanger. This last scheme is usually less expensive in implementation since it does not require an orifice with its associated flanges, which can be expensive. Both cascade schemes are common in the process industries. Can the reader say which of the two schemes gives a better initial response to disturbances in inlet process temperature $T_i(t)$?

The cascade control systems shown in Fig. 4-3.2*a* and *b* are very common in industrial practice. A typical application is in distillation columns where temperature is controlled to maintain the desired split. The temperature controller is often cascaded to the steam flow to the reboiler or the coolant flow to the condenser.

Finally, another very simple example of a cascade control system is that of a positioner on a control valve. The positioner acts as the inner controller of the cascade scheme.

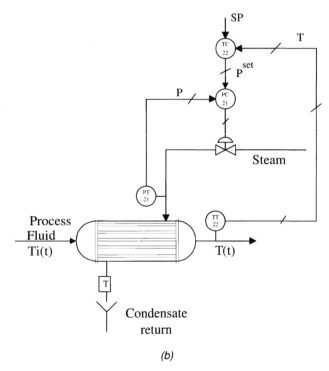

(b)

Figure 4-3.2   *Continued.*

4-4 CLOSING COMMENTS

So far, no comments have been made regarding the action of the controllers in a cascade strategy. This is important because, as learned in Chapter 3, if the actions are not chosen correctly, the controllers will not control. The procedure to choose the action is the same as explained in Chapter 3. That is, the action is decided by process requirements and the fail-safe action of the final control element. As noted previously, for some of the controllers in the cascade strategy, the final control element is the set point of another controller.

Consider the three-level cascade strategy shown in Fig. 4-3.1. The action of FC103 is reverse (Inc/Dec), because if the flow measurement increases above the set point, indicating that more flow than required is being delivered by the valve, the valve opening must be reduced, and for a fail-closed valve this is accomplished by reducing the signal to it. The action of TC102 is also reversed because if its measurement increases above the set point, indicating a higher outlet preheater temperature than required, the fuel flow must be reduced, and this is accomplished by reducing the set point to FC102. Finally, the action of TC101 is also reversed because if its measurement increases above the set point, indicating a higher reactor temperature than required, the way to reduce it is by lowering the inlet reactant's temperature, which is accomplished by reducing the set point to TC102. The decision regarding the controller action is simple and easy as long as we understand the significance of what each controller is doing.

Considering Fig. 4-2.1, the output from TC101 is a signal, meaning 4 to 20 mA or 3 to 15 psig or, in general, 0 to 100%. Then for a given output signal from TC101, say 40%, what is the temperature, in degrees, required from TC102? This question is easy to answer by remembering that the job of the controller is to make its measurement equal to the set point. Therefore, TC102 will be satisfied when the signal from TT102 is 40%. Thus the required temperature is 40% of the range of TT102.

Considering Fig. 4-2.1 again, it is important to realize what would happen if TC102 were taken off remote set-point operation while leaving TC101 in automatic. If this is done, and if TC101 senses an error, it would send a new signal (set point) to TC102. However, TC102 would be unable to respond to requests from TC101. If TC101 has reset action, it would wind up, since its output would have no effect in its input. That is, the effect of taking the secondary controller off remote set point is to "open" the feedback loop of the primary controller.

With their inherit flexibility, computers offer the necessary capabilities to avoid this windup possibility and thus provide for a safer cascade strategy. The computer can be programmed, or configured, so that at any time the secondary controller is taken off remote set-point operation, the primary controller "automatically" goes into manual mode if it is in automatic. The primary controller remains in manual as long as the secondary controller remains off remote set point. When the secondary controller is returned to remote set point, the primary controller could then return "automatically" to the automatic mode if the designer desires it. However, if while the secondary controller is off remote set point, its set point changes, then at the moment it is returned to remote set point mode, its present set point may not be equal to the output of the primary controller. If this occurs, the set point of the secondary controller will immediately jump to equal the output of the primary controller, thus generating a "bump" in the process operation. If a bumpless transfer is desired, most computer-based controllers can also be programmed so that while the secondary controller is off remote set point, the output from the primary controller is forced to equal either the process variable or the set point of the secondary controller. That is, the output from the primary controller "tracks" either variable of the secondary controller. Thus, when the secondary controller is returned to remote set point operation, a smooth transfer is obtained.

The tracking option just explained, often referred as *output tracking*, *reset feedback* (RFB), or *external reset feedback*, is very important for the smooth and safe operation of cascade control systems. We represent this option by the dashed lines in Fig. 4-2.1.

4-5 SUMMARY

In this chapter we have presented in detail the fundamentals and benefits of cascade control, which is a simple strategy, in concept and implementation, that provides improved control performance. The reader must remember that the secondary variable must respond faster to changes in the manipulated variable than the primary variable. Typical two-level cascaded loops are temperature to flow, concentration to flow, pressure to flow, level to flow, and temperature to pressure.

REFERENCES

1. G. Pressler, *Regelungs-Technik*, Hochschultashenbucher, Band 63, Bibliographischer Institut, Mannheim, Germany.
2. V. D. Austin, Development of tuning relations for cascade control systems, Ph.D. dissertation, Department of Chemical Engineering, University of South Florida, Tampa, FL, 1986.
3. A. B. Corripio, *Tuning of Industrial Control Systems*, Instrument Society of America, Research Triangle Park, NC, 1990.

CHAPTER 5

RATIO, OVERRIDE, AND SELECTIVE CONTROL

In Chapter 4 we began the presentation of control techniques that aid simple feedback to provide improved control performance. Specifically, in Chapter 4 we presented cascade control. In the present chapter we continue this presentation with three other techniques: *ratio*, *override*, and *selective control*; override control is also sometimes referred to as *constraint control*. Ratio control is commonly used to maintain two or more streams in a prescribed ratio. Override and selective control are usually implemented for safety and optimization considerations. These two techniques often deal with multiple control objectives (controlled variables) and a single manipulated variable; up to now we have dealt only with processes with one control objective. The chapter begins with a presentation of distributed control systems (DCSs), how they handle signals, and some computing algorithms and programming needed for implementing control techniques.

5-1 SIGNALS AND COMPUTING ALGORITHMS

Many of the control techniques presented in this and subsequent chapters require some amount of computing power. That is, many of these techniques require the multiplication, division, addition, subtraction, and so on, of different signals. Several years ago all of these calculations were implemented with analog instrumentation. Computers allow for a simpler, more flexible, more accurate, more reliable, and less expensive implementation of these functions.

5-1.1 Signals

There are two different ways that field signals are handled once they enter the DCS. The first way is to convert the signal received by the computer into a number with engineering units. For example, if a signal is read from a temperature transmitter,

74

the number kept in memory by the computer is the temperature in degrees. The computer is given the low value of the range and the span of the transmitter, and with this information it converts the raw signal from the field into a number in engineering units. A possible command in the DCS to read a certain input is

variable = AIN(input channel #, low value of range, span of stransmitter)

or

$$T = \text{AIN}(3, 50, 100)$$

This command instructs the DCS to read an analog input signal (AIN) in channel 3, it tells the DCS that the signal comes from a transmitter with a low value of 50 and a span of 100, and it instructs the DCS to assign the name T to the variable read (possibly a temperature from a transmitter with a range of 50 to 150°C). If the signal read had been 60%, 13.6 mA, then $T = 110°C$.

The second way of handling signals, and fortunately the least common, is not by converting them to engineering units but by keeping them as a percentage, or fraction, of the span. In this case the input command is something like

variable = AIN(input channel)

or

$$T = \text{AIN}(3)$$

and the result, for the same example, is $T = 60\%$ (or 0.6).

In DCSs that work in engineering units, the range of the transmitter providing the controlled variable must be supplied to the PID controller (there are different ways to do so). With this information, the controller converts both the variable and the set point to percent values before applying the PID algorithm. This is done because the error is calculated in %TO. Remember, the K_C units are %CO/%TO. Thus the controller output is then %CO. A possible way to "call" a PID controller could be

OUT = PID(controlled variable, set point, low value of range, span of transmitter)

or

$$\text{OUT} = \text{PID}(T, 75, 50, 100)$$

This command instructs the DCS to control a variable T at 75 (degrees) that is supplied by a transmitter with a range from 50 to 150 (degrees). The controller output (OUT) is in percent (%CO).

5-1.2 Programming

There are two ways to program the mathematical manipulations in DCSs: block-oriented programming and software-oriented programming.

Block-Oriented Programming. Block-oriented programming is software in a subroutine-type form, referred to as computing *algorithms* or computing *blocks*. Each block performs a specified mathematical manipulation. Thus, to develop a control strategy, the computing blocks are linked together, the output of one block being the input to another block. This linking procedure is often referred to as *configuring* the control system.

Some typical calculations (there are many others) performed by computing blocks are:

1. *Addition/subtraction.* The output signal is obtained by adding and/or subtracting the input signals.
2. *Multiplication/division.* The output signal is obtained by multiplying and/or dividing the input signals.
3. *Square root.* The output signal is obtained by extracting the square root of the input signal.
4. *High/low selector.* The output signal is the highest/lowest of two or more input signals.
5. *High/low limiter.* The output signal is the input signal limited to a preset high/low limit value.
6. *Function generator, or signal characterization.* The output signal is a function of the input signal. The function is defined by configuring the x, y coordinates.
7. *Integrator.* The output signal is the time integral of the input signal. The industrial term for *integrator* is *totalizer*.
8. *Lead/lag.* The output signal is the response of the transfer function given below. This calculation is often used in control schemes, such as feedforward, where dynamic compensation is required.

$$\text{Output} = \frac{\tau_{\text{ld}} s + 1}{\tau_{\text{lg}} s + 1} \cdot \text{input}$$

9. *Dead time.* The output signal is equal to a delayed input signal. This calculation is very easily done with computers but is extremely difficult to do with analog instrumentation.

Table 5-1.1 shows the notation and algorithms we use in this book for mathematical calculations. Often, these blocks are linked together graphically using standard "drag-and-drop" technology.

Software-Oriented Programming. Manufacturers have developed their own programming languages, but they are all similar and resemble Fortran, Basic, or C. Table 5-1.2 presents the programming language we use in this book; this language is similar to those used by different manufacturers.

5-1.3 Scaling Computing Algorithms

When signals are handled as a percent, or fraction, of span, additional calculations must be performed before the required mathematical manipulations can be

TABLE 5-1.1   Computing Blocks

OUT = output from block
I_1, I_2, I_3 = input to blocks
K_0, K_1, K_2, K_3 = constants that are used to multiply each input
B_0, B_1, B_2, B_3 = constants

Summer: $\text{OUT} = K_1 I_1 + K_2 I_2 + K_3 I_3 + B_0$

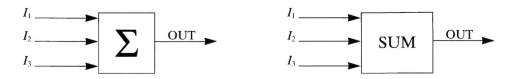

Multiplier/divider: $\text{OUT} = \dfrac{K_0 (K_1 I_1 + B_1)(K_2 I_2 + B_2)}{K_3 I_3 + B_3} + B_0$

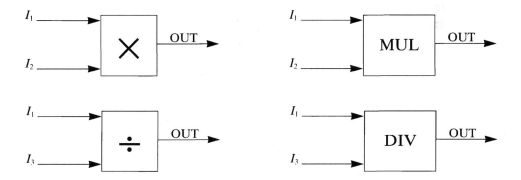

Square root: $\text{OUT} = K_1 \sqrt{I_1}$

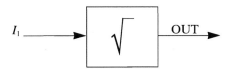

Lead/lag: $\text{OUT} = \dfrac{K_0 (\tau_{ld} s + 1)}{\tau_{lg} s_1 + 1} I_1$

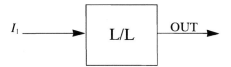

(Continued)

TABLE 5-1.1 *Continued*

Selector: OUT = maximum of inputs I_1, I_2, I_3
OUT = minimum of inputs I_1, I_2, I_3

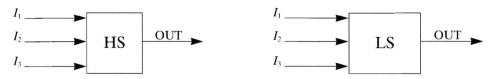

Dead time: OUT = input delayed by t_0

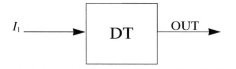

TABLE 5-1.2 Programming Language

Input/output: AIN = analog in; AOUT = analog out
Format:

    In variable = AIN (input channel #, low value of range, span of transmitter)
    "In variable" will be returned in engineering units.
    Out variable = AOUT (output channel #, out variable)
    "Out variable" will be returned in percent.

Mathematical symbols: +, −, *, ^, /, <, >, =

Statements: GOTO; IF/THEN/ELSE

Controller:
    Output = PID (variable, set point, low value of range of variable, span of variable)
    "Output" will be returned in percent.
    Every term in the PID argument must be in engineering units.

Comments: To insert a comment in any line, use a semicolon followed by the comment.

---

implemented. The necessity and meaning of the additional calculations are explained by the following. Consider a tank, shown in Fig. 5-1.1, where temperature transmitters with different ranges measure temperatures at three different locations in the tank. The figure shows the transmitter ranges and the steady-state values of each temperature, which are at midvalue of each range. It is desired to compute the average temperature in the tank. This computation is straightforward for the control system that reads each signal and converts it to engineering units. The three values are added together and divided by 3; the program in Fig. 5-1.2 does just that. The first three lines, T101, T102, and T103, read in the temperature, and the fourth statement calculates the average temperature, TAVG.

For control systems that treat each signal as a percent of span, this simple com-

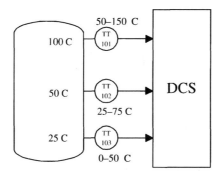

Figure 5-1.1   Tank with three temperature transmitters.

1	T101=AIN(1,50,100) ; reads in T101
2	T102=AIN(2,25,50) ; reads in T102
3	T103=AIN(3,0,50) ; reads in T103
4	TAVG=(T101+T102+T103)/3 ; calculates average

Figure 5-1.2   Program to read in temperatures, in engineering units, and calculate average temperature.

1	T101=AIN(1) ; reads in T101
2	T102=AIN(2) ; reads in T102
3	T103=AIN(3) ; reads in T103
4	TAVG=(T101+T102+T103)/3 ; calculates average

Figure 5-1.3   Program to read in temperatures, in percent of span, and calculate average temperature.

putation would result in an answer without much significance; Fig. 5-1.3 shows this program. That is, because each signal is 50% of its range, the computation result would also be 50%. However, 50% of what range? How do we translate this answer into a temperature? Furthermore, notice that even though every input signal is 50%, their measured temperatures are different because the ranges are different. Thus, for the computation to "make sense," the range of each input signal, and a chosen range for the output variable, must be considered. The consideration of each range will ensure compatibility between input and output signals, and it is called *scaling*. Reference 1 presents the method to scale the computations.

5-1.4   Significance of Signals

During the presentation of the types of field signals in Chapters 1 and 4, and in the discussion earlier in this section, it was mentioned that signals are used by the instruments to convey information and that, therefore, every signal has *physical significance*; that is, every signal used in the control scheme has some meaning. Signals are in percent, but percent of what (pressure, temperature, flow, etc.)? The *what* is the meaning of the signal. It is now important to stress this fact again as we embark on the design of complex strategies to improve control performance.

As mentioned earlier in this chapter, the new strategies frequently require the manipulation of signals in order to calculate controlled variables, set points, or decide on control actions. To perform these calculations correctly, it is most important to understand the significance of the signals.

Very often, the first step in the design of a control strategy is to give a signal, sometimes referred to as the *master signal*, a physical significance. Then, based on the given significance, the strategy is designed. Currently, this presentation may seem somewhat abstract; however, as we continue with the study of different control strategies, the presentation will become clear and realistic.

To help keep all the information in order and to understand the calculations, we indicate next to each signal its significance and direction of information flow. This practice is not common in industry, but it helps in learning and understanding the subject.

5-2 RATIO CONTROL

A commonly used process control technique is *ratio control*, which is the term used to describe the strategy where one variable is manipulated to keep it as a ratio or proportion of another. In this section we present two industrial examples to show its meaning and implementation. The first example is a simple and common one and explains clearly the need for ratio control.

Example 5-2.1. Assume that it is required to blend two liquid streams, A and B, in some proportion, or ratio, R; the process is shown in Fig. 5-2.1. The ratio is $R = F_B/F_A$, where F_A and F_B are the flow rates of streams A and B, respectively.

An easy way of accomplishing this task is shown in the figure. Each stream is controlled by a flow loop in which the set points to the controllers are set such that

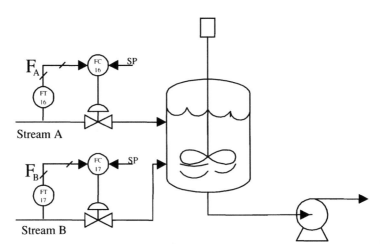

Figure 5-2.1 Control of blending of two liquid streams.

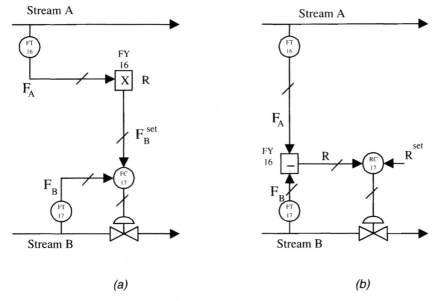

Figure 5-2.2 Ratio control of blending system.

the liquids are blended in the correct ratio. However, suppose now that one of the streams, stream A for example, cannot be controlled, just measured. The flow rate of this stream, often referred to as *wild flow*, is usually manipulated to control something else, such as level or temperature, upstream. The controlling task is now more difficult. Somehow the flow rate of stream B must vary, as the flow rate of stream A varies, to maintain the blend in the correct ratio. Two possible ratio control schemes are shown in Fig. 5-2.2.

The scheme shown in Fig. 5-2.2a consists of measuring the wild flow and multiplying it by the desired ratio, in FY16, to obtain the required flow rate of stream B; that is, $F_B^{set} = RF_A$. The output of FY16 is the required flow rate of stream B, and it is used as the set point to the flow controller of stream B, FC17. So as the flow rate of stream A varies, the set point to the flow controller of stream B will vary accordingly to maintain both streams at the ratio required. If a new ratio between the two streams is required, the new ratio is set in the multiplier. The set point to the flow controller of stream B is set from a computation, not locally. Figure 5-2.3a shows the software equivalent to Fig. 5-2.2a and assumes that the control system works in engineering units. FT16LO, FT16SPAN, FT17LO, and FT17SPAN are the low value and span of FT16 and FT17.

The ratio control scheme shown in Fig. 5-2.2b consists of measuring both streams and dividing them, in FY16, to obtain the actual ratio flowing through the system. The calculated ratio is then sent to a controller, RC17, which manipulates the flow of stream B to maintain the set point. The set point to this controller is the required ratio and it is set locally. Figure 5-2.3b shows the equivalent scheme using software. Note that in the controller it is necessary to specify RLO and RSPAN, which are the low value and span you expect the ratio to change. This is the same as selecting a ratio transmitter range.

```
1       FA=AIN(1, FT16LO, FT16SPAN) ; reads in flow of stream A
2       FB=AIN(2, FT17LO, FT17SPAN) ; reads in flow of stream B
3       FBSET=R*FA ; FY16
4       CO17=PID(FB, FBSET, FT17LO, FT17SPAN) ; FC17
5       AOUT(1, CO17) ; outputs signal to valve
```

(a)

```
1       FA=AIN(1, FT16LO, FT16SPAN) ; reads in flow of stream A
2       FB=AIN(2, FT17LO, FT17SPAN) ; reads in flow of stream B
3       RCALC=FB/FA ; FY16
4       CO17=PID(RCALC, R, RLO, RSPAN) ; RC17
5       AOUT(1, CO17)
```

(b)

Figure 5-2.3   Software equivalent of Fig. 5-2.2.

Both control schemes shown in Fig. 5-2.2 are used, but the scheme shown in Fig. 5-2.2a is preferred because it results in a more linear system than the one shown in Fig. 5-2.2b. This is demonstrated by analyzing the mathematical manipulations in both schemes. In the first scheme FY16 solves the equation $F_B^{set} = RF_A$. The gain of this device, that is, how much its output changes per change in flow rate of stream A, is given by

$$\frac{\partial F_B^{set}}{\partial F_A} = R$$

which is a constant value. In the second scheme, FY16 solves the equation

$$R = \frac{F_B}{F_A}$$

Its gain is given by

$$\frac{\partial R}{\partial F_A} = \frac{F_B}{F_A^2} = \frac{R}{F_A}$$

so as the flow rate of stream A changes, this gain also changes, yielding a nonlinearity.

From a practical point of view, even if both streams can be controlled, the implementation of ratio control may still be more convenient than the control system shown in Fig. 5-2.1. Figure 5-2.4 shows a ratio control scheme for this case. If the total flow must be changed, the operator needs to change only one flow, the set point

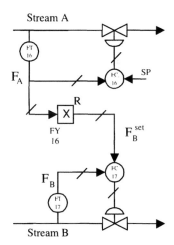

Figure 5-2.4 Ratio control of blending system.

to FC16; then the set point to FC17 changes automatically once the flow rate of stream A changes. In the control system of Fig. 5-2.1 the operator needs to change two flows, the set points to FC16 and FC17.

The schemes shown in Figs. 5-2.2a and 5-2.4 are quite common in the process industries. Recalling what was presented about computing blocks in section 5-1, we realize that the implementation of the ratio stations can simply be accomplished with the use of a unit such as the one shown in Table 5-1.2. Most computer control systems offer a controller, referred to as PID-RATIO, that accepts a signal, applies the same algorithm as the ratio unit, FY16, in Fig. 5-2.2a, and uses the internal result as its set point. Thus, if a PID-RATIO is used, the calculations done by FY16 and FC17 in Fig. 5-2.4 are performed in only one block.

As we have mentioned several times already, it is helpful in developing control schemes to remember that every signal must have physical significance. In Figs. 5-2.2 and 5-2.4 we have labeled each signal with its significance. For example, in Fig. 5-2.2a the output signal from FT16 is related to the flow rate of stream A and has the label F_A. If this signal is then multiplied by the ratio F_B/F_A, or simply R, the output signal from FY16 is the required flow rate of stream B, F_B^{set}. Even though this use of labels is not standard practice, for pedagogical reasons we continue to label signals with their significance throughout the chapter. We recommend that the reader do the same.

Example 5-2.2. Another common example of ratio control used in the process industries is control of the air/fuel ratio to a boiler or furnace. Air is introduced in a set excess of that required stoichiometrically for combustion of the fuel; this is done to ensure complete combustion. Incomplete combustion results not only in inefficient use of the fuel, but may also result in smoke and the production of pollutants. In addition, if not enough air is introduced, this may result in pockets of pure fuel inside the combustion chamber—not a very safe condition. The excess air introduced is dependent on the type of fuel, fuel composition, and equipment used. However, the greater the amount of excess air introduced, the greater the energy

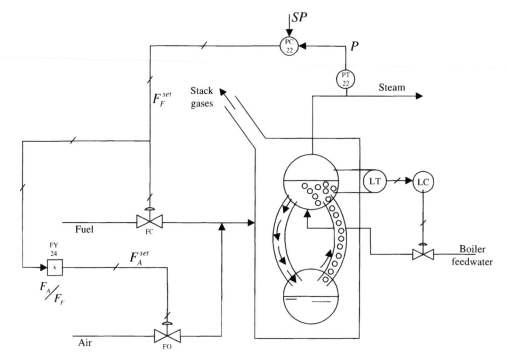

Figure 5-2.5 Parallel positioning control with manually adjusted air/fuel ratio.

losses through the stack gases. Therefore, control of the air and fuel flows is most important for safe and economical operation.

The flow of combustibles is generally used as the manipulated variable to maintain the pressure of the steam produced in the boiler at some desired value. Figure 5-2.5 shows one way to control the steam pressure as well as the air/fuel ratio control scheme. This scheme is called *parallel positioning control* [2–4] with manually adjusted fuel/air ratio. The steam pressure is transmitted by PT22 to the pressure controller PC22, and this controller manipulates a signal, often referred to as the *boiler master signal*, to the fuel valve. Simultaneously, the controller also manipulates the air damper through the ratio unit FY24. This ratio station sets the air/fuel ratio required.

The control scheme shown in Fig. 5-2.5 does not actually maintain an airflow/fuel flow ratio, but rather, maintains only a ratio of signals to the final control elements; the actual flows are not measured and used. The flow through the valves depends on the signals and on the pressure drop across them. Consequently, any pressure fluctuation across the valve or air damper changes the flow, even though the opening has not changed, and this in turn affects the combustion process and steam pressure. A better control scheme to avoid this type of disturbance, shown in Fig. 5-2.6, is referred to as *full metering control* [2]. (Figure 5-2.6 is referred to as a top-down instrumentation diagram, and it is commonly used to present control schemes.) In this scheme the pressure controller sets the flow of fuel, and the airflow is ratioed from the fuel flow. The flow loops correct for any flow disturbances. The fuel/air ratio is still adjusted manually.

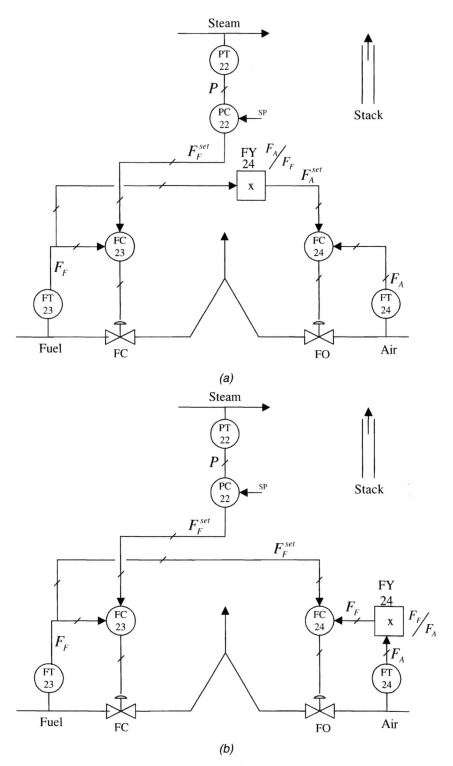

(a)

(b)

Figure 5-2.6 Full metering control with manually adjusted fuel/air ratio.

Notice the differences between the two figures. In Fig. 5-2.6a the signal from FT23 is multiplied by the ratio F_F/F_A before it is used as the set point to FC24; note that the significance of all signals make sense. Figure 5-2.6b is the one that seems somewhat strange. The figure shows that the signal setting the set point to FC24 comes from FT23; therefore, it is related to F_F; FC24 is the controller that moves the airflow. Note, however, that the signal from FT24, which is related to the airflow, is multiplied by F_F/F_A before it is used as the measurement to FC24. Thus, both the measurement and the set point to FC24 have the same meaning. It seems that Fig. 5-2.6b is somewhat more difficult to understand, but its use in the following schemes results in fewer blocks to use.

Let us analyze the control scheme shown in Fig. 5-2.6 in more detail. When the steam header pressure increases, probably due to a decrease in steam demand, the pressure controller reduces the demand for fuel. As the set point to the fuel flow controller is reduced, the controller closes the valve to satisfy the set point. As the fuel flow decreases, the set point to the airflow controller is also reduced. Thus the airflow follows the fuel flow, and during a transient period the entering combustible mixture is richer in air than usual. Let us now consider the case when the header pressure decreases, probably due to an increase in steam demand, and the pressure controller increases the demand for fuel. As the set point to the fuel flow controller increases, the controller opens the valve to satisfy the set point. As the fuel flow increases, the set point to the airflow then increases; the airflow again follows the fuel flow. In this last case, during a transient period, the entering combustible mixture is not richer in air, and if not careful, it may even be lean in air. This situation is certainly not desirable, for the reasons explained at the beginning of the example. Therefore, a control scheme must be designed to avoid these situations. The control scheme must be such that when more combustibles are required to maintain pressure, it increases the air first, followed by the fuel. When fewer combustibles are required, it decreases the fuel first, followed by the air. This pattern ensures that during transient periods the combustible mixture is air-rich. Figure 5-2.7 shows a scheme, referred to as *cross-limiting control*, that provides the required control. Only two selectors, LS23 and HS24, are added to the previous control scheme. The selectors provide a way to decide which device sets the set point to the controller. The reader is encouraged to go through the scheme to understand how it works. As a way to do so, assume that the required air/fuel ratio is 2 and that at steady state the required fuel is 10 units of flow. Consider next what happens if the header pressure increases and the pressure controller asks for only 8 units of fuel flow. Finally, consider what happens if the header pressure decreases and the pressure controller asks for 12 units of fuel flow.

Since the amount of excess air is important for economical, environmentally sound operation of the boilers, it has been proposed to provide a feedback signal based on an analysis of the stack gases; the analysis is often percent O_2 or percent CO. Based on this analysis, it is then proposed that the fuel/air ratio be adjusted. This new scheme shown in Fig. 5-2.8 consists of an analyzer transmitter, AT25, and a controller, AC25. The controller maintains the required percent O_2, for example, in the stack gases by setting the required fuel/air ratio.

Before finishing this section it is interesting to see how the control scheme shown in Fig. 5-2.8 is programmed using the software language; this is presented in

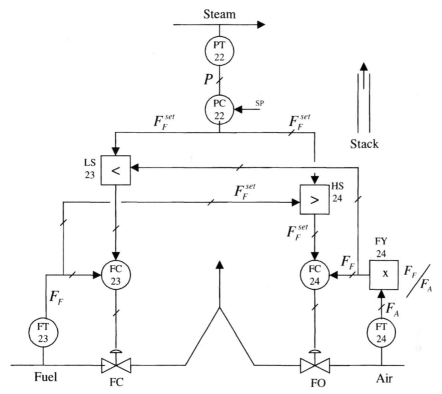

Figure 5-2.7 Cross-limiting control.

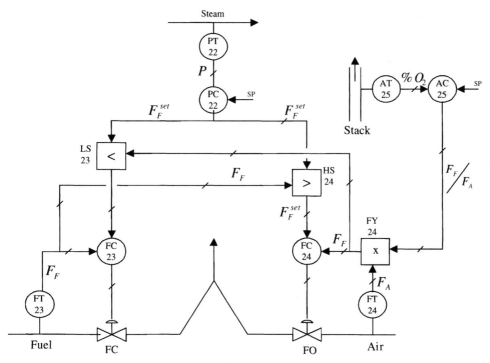

Figure 5-2.8 Cross-limiting with O_2 trim control.

1     $P = AIN(1, P_{low}, P_{span})$; reads in pressure

2     $FA = AIN(2, FA_{low}, FA_{span})$; reads in air flow

3     $FF = AIN(3, FF_{low}, FF_{span})$; reads in fuel flow

4     $\%O2 = AIN(4, \%O2_{low}, \%O2_{span})$; reads in $\%O_2$

5     $FOUT = PID(P, P^{set}, P_{low}, P_{span})$; PC22

6     $ROUT = PID(\%O2, \%O2^{set}, \%O2_{low}, \%O2_{span})$; AC25

7     $PFF^{set} = (FF_{span}/100)*FOUT + FF_{low}$; converts output of PC22 to
                                    ; fuel flow set point in engineering units

8     $RATIO = (RATIO_{span}/100)*ROUT + RATIO_{low}$; converts output
                                       ;of AC25 to FA/FF ratio in engineering units

9     $RFF = FA*RATIO$; FY24

10   IF $PFF^{set} < RFF$ THEN $FF^{set} = PFF^{set}$ ELSE $FF^{set} = RFF$; LS23

11    $COFUEL = PID(FF, FF^{set}, FF_{low}, FF_{span})$; FC23

12   IF $PFF^{set} > FF$ THEN $FF^{set} = PFF^{set}$ ELSE $FF^{set} = FF$; HS24

13    $COAIR = PID(RFF, FF^{set}, FF_{low}, FF_{span})$; FC24

14    $AOUT(1, COFUEL)$; outputs signal to fuel valve

15    $AOUT(2, COAIR)$; outputs signal to air valve

Figure 5-2.9 Software program equivalent to Fig. 5-2.8.

Fig. 5-2.9. The comments associated with each statement help to relate the program to Fig. 5-2.8. In this section we have shown two applications of ratio control. As mentioned at the beginning of the section, ratio control is a common technique used in the process industries; it is simple and easy to use.

5-3 OVERRIDE, OR CONSTRAINT, CONTROL

Override, or *constraint, control* is a powerful yet simple control strategy generally used as a protective strategy to maintain process variables within limits that must be enforced to ensure the safety of personnel and equipment and product quality. As a protective strategy, override control is not as drastic as interlock control. Interlock controls are used primarily to protect against equipment malfunction. When a malfunction is detected, the interlock system usually shuts the process down. Interlock systems are not presented, but Refs. 5 and 6 are provided for their study. Two examples of constraint control are now presented to demonstrate the concept and implementation of the strategy.

Example 5-3.1. Consider the process shown in Fig. 5-3.1. A hot saturated liquid enters a tank and from there is pumped under flow control back to the process. Under normal operation the level in the tank is at height h_1. If, under any circumstances, the liquid level drops below the height h_2, the liquid will not have enough net positive suction head (NPSH), and cavitation at the pump will result. It is therefore necessary to design a control scheme that avoids this condition. This new control scheme is shown in Fig. 5-3.2.

The level in the tank is now measured and controlled. The set point to LC50 is somewhat above h_2, as shown in the figure. It is important to notice the action of the controllers and final control element. The variable-speed pump is such that as

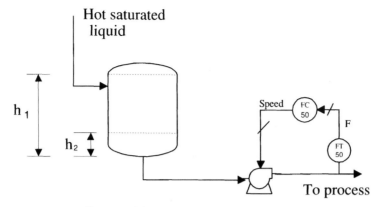

Figure 5-3.1   Tank and flow control loop.

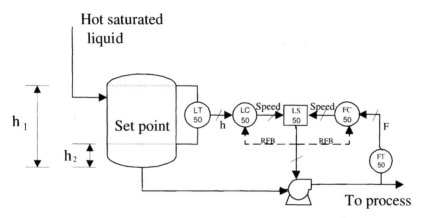

Figure 5-3.2   Override control scheme.

the input energy (current in this case) to it increases, it pumps more liquid. Therefore, the FC50 is a reverse-acting controller, while the LC50 is a direct-acting controller. The output of each controller is connected to a low selector, LS50, and the
signal from this selector goes to the pump.

Under normal operating conditions the level is at h_1, which is above the set point
to the level controller; consequently, the controller will try to speed up the pump as
much as possible, increasing its output to 100%. Under normal conditions the output
of the flow controller may be 75%, and the low selector selects this signal to manipulate the pump speed. Thus, under normal conditions the flow controller is manipulating the pump. The level controller is not connected to the pump because the
level is not at an undesirable state. This is the desired operating condition.

Let us now suppose that the flow of hot saturated liquid into the tank slows down
and the level starts to drop. As soon as the level drops below the set point on the
level controller, the controller will try to slow down the pump by reducing its output.
When the level controller's output drops below the output of the flow controller,
the low selector selects the output of the level controller to manipulate the pump.
It can be said that the level controller "overrides" the flow controller.

When the flow of hot liquid returns to its normal value, and the level increases above the set point, the level controller increases its output to speed up the pump. Once the output from the level controller increases above the output from the flow controller, the low selector selects the flow controller, and operation returns to its normal condition.

An important consideration in designing an override control system is that of reset windup protection on any controller that has integration. The output of the controller not selected must stop at 100%, not at a higher value, or at 0%, not at a lower value. Even more desirable would be an operation in which if the controller output selected were 75%, the nonselected output would be forced to be close to 75%. With their inherent flexibility, DCSs provide this very important capability. What is done is that the integration portion of the controller(s) not selected is (are) stopped and forced equal to the output of the selector. For example, under normal operating conditions the low selector, LS50, selects the flow controller and not the level controller. In this case the integration of the level controller is stopped and forced to equal the output of LS50 (e.g., 75%). Mathematically, we can explain this procedure by looking at the PI equation of the level controller,

$$m(t) = K_C e(t) + \frac{K_C}{\tau_I} \int e(t)\,dt = K_C e(t) + \text{LS}50(75\%)$$

Furthermore, knowing that $(K_C/\tau_I)\int e(t)\,dt = \text{LS}50\ (75\%)$, and the tuning parameters K_C and τ_I, the value of the integral $\int e(t)\,dt$ is back-calculated continuously. The proportional part of the controller is allowed to continue working. What this accomplishes is that under normal operating conditions, the output of the level controller is greater than that of the flow controller, because the proportional part is positive and it adds to the integral part, which is kept at a value equal to LS50. However, at the moment the level in the tank is equal to the set point of the controller, the error is zero, and the level controller output is equal to the output of the LS and therefore equal to the output of the flow controller. As soon as the level in the tank drops below the set point, the term $K_C e(t)$ becomes negative and the output of the level controller is less than that of the flow controller, and thus is selected by the low selector. At that moment the integral term of the level controller is permitted to start integrating again, starting from the last value from which it was back calculated.

This capability is referred to as *reset feedback* (RFB), or sometimes as *external reset feedback*. We use the dashed lines shown in Fig. 5-3.2 to indicate that the controller is using this capability. The figure shows the RFB capability to both controllers. When FC50 is being selected, its integration is working, but not that of LC50 (its integration is being forced equal to the output of LS50). When LC50 is being selected, its integration is working but not that of FC50 (its integration is being forced equal to the output of LS50). The selection of this capability is very easily done in DCSs, and once selected, all the calculations just explained are transparent to the user.

To summarize, the reset feedback capability allows the controller not selected to override the controller selected *at the very moment it is necessary*. More than two controllers can provide signals to a selector and have RFB signals; this is shown in the following example.

Example 5-3.2. A fired heater, or furnace, is another common process that requires the implementation of constraint control. Figure 5-3.3 shows a heater with temperature control manipulating the gas fuel flow. The manipulation of the combustion air has been omitted to simplify the diagram; however, it is the same as discussed in detail in Section 5-2. There are several conditions in this heater that can prove quite hazardous. Some of these conditions are higher fuel pressure, which can sustain a stable flame, and higher stack, or tube, temperature than the equipment can safely handle. If either of these conditions exist, the gas fuel flow must decrease to avoid the unsafe condition; at this moment, temperature control is certainly not as important as the safety of the operation. Only when the unsafe conditions disappear is it permissible to return to straight temperature control.

Figure 5-3.4 shows a constraint control strategy to guard against the unsafe condition described above. The gas fuel pressure is usually below the set point to PC14, and consequently, the controller will try to raise the set point to the fuel flow controller. The stack temperature will also usually be below the set point to TC13, and consequently, the controller will try to raise the set point to the fuel flow controller. Thus, under normal conditions the exit heater temperature controller would be the controller selected by the low selector because its output will be the lowest of the three controllers. Only when one of the unsafe conditions exist would TC12 be "overridden" by one of the other controllers.

As explained in Example 5-3.1, it is important to prevent windup of the controllers that are not selected. Thus the control system must be configured, or

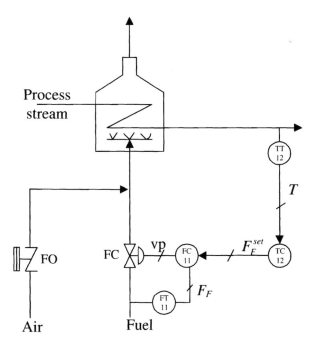

Figure 5-3.3   Heater temperature control.

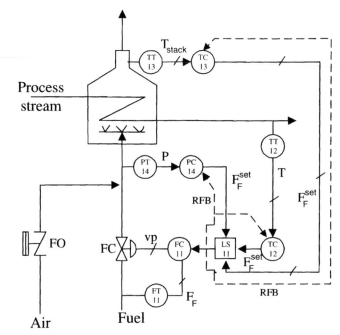

Figure 5-3.4 Heater temperature control, constraint control.

programmed, to provide reset feedback. This is shown by the dashed lines in the figure.

The constraint control scheme shown in Fig. 5-3.4 contains a possible safety difficulty. If at any time the operating personnel were to set the flow controller FC11 in local set point or in the manual mode (i.e., off remote set point), the safety provided by TC13 and PC14 would not be in effect. This would result in an unsafe and unacceptable operating condition. You may want to think how to design a new constraint control strategy to permit the operating personnel to set the flow controller in automatic or manual and still have the safety provided by TC13 and PC14 in effect.

The introduction to this section mentioned that override control is commonly used as a protective scheme. Examples 5-3.1 and 5-3.2 presented two of these applications. As soon as the process returns to normal operating conditions, the override scheme returns automatically to its normal operating status. The two examples presented show multiple control objectives (controlled variables) with a single manipulated variable; however, only one objective is enforced at a time.

5-4 SELECTIVE CONTROL

Selective control is another interesting control scheme used for safety considerations and process optimization. Two examples are presented to show its principles and implementation.

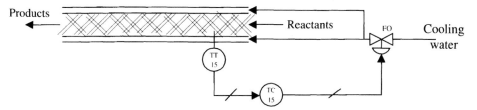

Figure 5-4.1 Temperature control of a plug flow reactor.

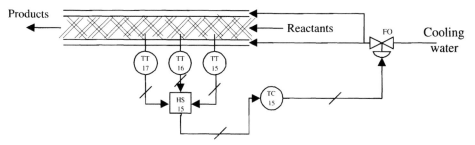

Figure 5-4.2 Selective control for a plug flow reactor.

Example 5-4.1. Figure 5-4.1 shows a plug flow reactor where an exothermic catalytic reaction takes place; the figure also shows the reactor temperature control. The sensor providing the temperature measurement should be located at the "hot spot." As the catalyst in the reactor ages, or conditions change, the hot spot will move. It is desired to design a control scheme so that its measured variable "moves" as the hot spot moves. A control strategy that accomplishes the desired specifications is shown in Fig. 5-4.2. The high selector in this scheme selects the transmitter with the highest output, and in so doing the controlled variable is always the highest, or closest to the highest, temperature.

In implementing this control strategy an important consideration is that all temperature transmitters must have the same range, so that their output signals can be compared on the same basis. Another possibly important consideration is to install some kind of indication as to which transmitter is giving the highest signal. If the hot spot moves past the last transmitter, TT17, this may be an indication that it is time either to regenerate or to change the catalyst. The length of reactor left for the reaction is probably not enough to obtain the conversion desired.

Example 5-4.2. An instructive and realistic process where selective control can improve the operation is shown in Fig. 5-4.3. A furnace heats a heat transfer oil to provide an energy source to several process units. Each individual unit manipulates the flow of oil required to maintain its controlled variable at set point. The outlet oil temperature from the furnace is also controlled by manipulating the fuel flow. A bypass control loop, DPC16, is provided.

Suppose that it is noticed that the control valve in each unit is not open very much. For example, suppose that the output of TC13 is only 20%, that of TC14 is 15%, and that of TC15 is only 30%. This indicates that the hot oil temperature provided by the furnace may be higher than required by the users. Consequently, not

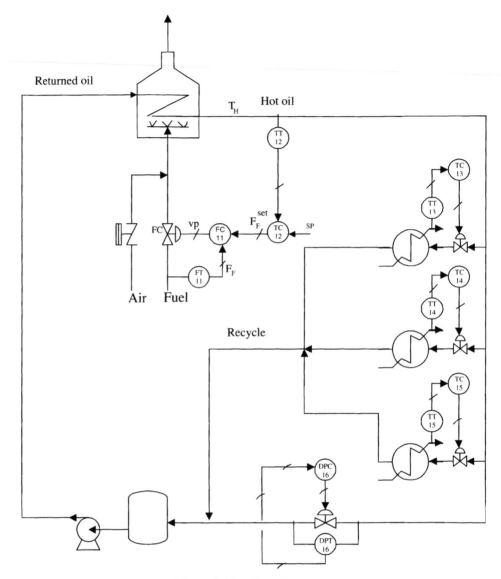

Figure 5-4.3   Hot oil system.

much oil flow is necessary and much of it will bypass the users. This situation is energy inefficient since to obtain a high oil temperature, a large quantity of fuel must be burned. Also, a significant amount of the energy provided by the fuel is lost to the surroundings in the piping system and through the stack gases.

A more efficient operation is the one that maintains the oil leaving the furnace at a temperature just hot enough to provide the necessary energy to the users, with hardly any flow through the bypass valve. In this case the temperature control valves would generally be open. Figure 5-4.4 shows a selective control strategy that provides this type of operation. The strategy first selects the most open valve using a high selector, HS16. The valve position controller, VPC16, controls the valve posi-

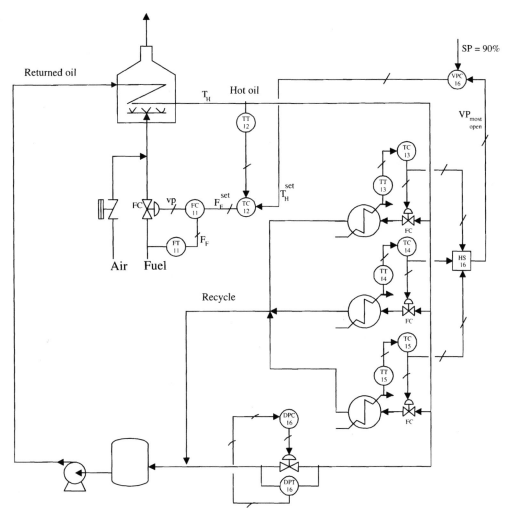

Figure 5-4.4 Selective control for hot oil system.

tion selected, say at 90% open, by manipulating the set point of the furnace temperature controller. Thus this strategy ensures that the oil temperature from the
furnace is just "hot enough."

Note that since the most open valve is selected by comparing the signals to each
valve, all the valves should have the same characteristics.

The selective control strategy shows again that with a bit of logic, a process operation can be improved significantly.

5-5 DESIGNING CONTROL SYSTEMS

In this section we present three examples to provide some hints on how to go about
designing control schemes. To obtain maximum benefit from this section, we

recommend that you first read the example statement and try to solve the problem by yourself. Then check with the solution presented.

Example 5-5.1. Consider the reactor shown in Fig. 5-5.1, where the exothermic reaction A + B → C takes place. The diagram shows the control of the temperature in the reactor by manipulating the cooling water valve.

(a) Design a control scheme to control the flow of reactants to the reactor. The flows of reactants A and B enter the reactor at a certain ratio R; that is, $R = F_B/F_A$. Both flows can be measured and controlled.

(b) Operating experience has shown that the inlet cooling water temperature varies somewhat. Because of the lags in the system, this disturbance usually results in cycling the temperature in the reactor. The engineer in charge of this unit has been wondering whether some other control scheme can help in improving the temperature control. Design a control scheme to help him.

(c) Operating experience has also shown that under infrequent conditions the cooling system does not provide enough cooling. In this case the only way to control the temperature is by reducing the flow of reactants. Design a control scheme to do this automatically. The scheme must be such that when the cooling capacity returns to normal, the scheme of part (b) is reestablished.

SOLUTION: (a) Figure 5-2.4 provides a scheme that can be used to satisfy the ratio control objective; Fig. 5-5.2 shows the application of the scheme to the present process. The operator sets the flow of stream A, set point to FC15, and the flow of stream B is set accordingly.

(b) A common procedure we follow to design control schemes is to first think what we would do to control the process manually. In the case at hand, after some thinking you may decide that it would be nice if somehow you be notified as soon as possible of a change in cooling water temperature. If this change is known, you could do something to negate its effect. For example, if the cooling water temperature increases, you could open the valve to feed in more fresh water; Fig. 5-5.3 shows this idea. But, you now think, I'm not considering the temperature controller

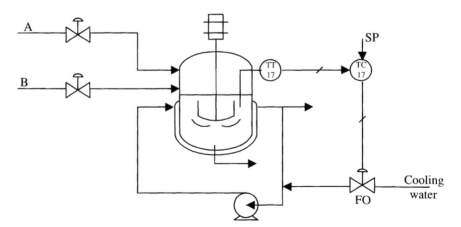

Figure 5-5.1   Reactor for Example 5-5.1.

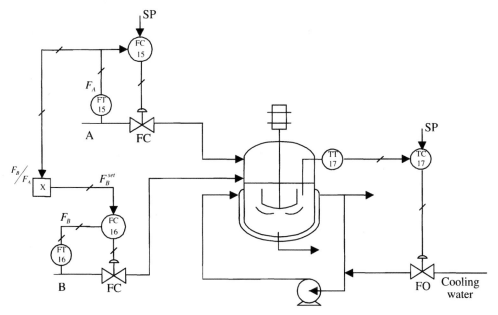

Figure 5-5.2   Ratio control scheme for part (a) of Example 5-5.1.

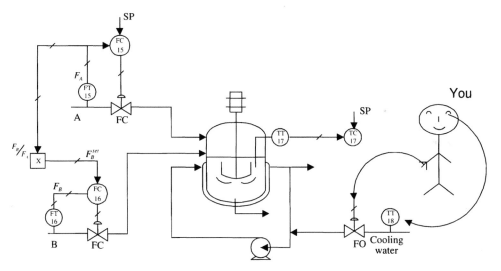

Figure 5-5.3   Proposed manual control scheme (first draft) to compensate for changes in inlet cooling water temperature.

TC17 at all. Well, why not use the output of TC17 as my set point, as a cascade control scheme; Fig. 5-5.4 shows this proposed scheme. Next, you decide to automate your idea, and for that you sketch Fig. 5-5.5. You have replaced yourself by another intelligence: a controller.

Now that you have sketched your idea, you need to analyze it further. The figure shows that the master controller, TC17, looks at the temperature in the reactor, com-

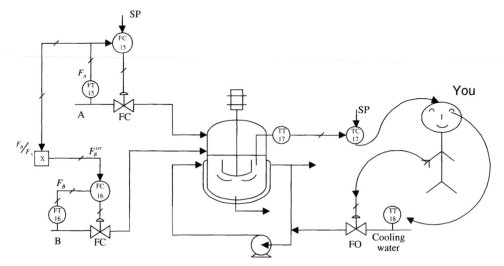

Figure 5-5.4   Proposed manual control scheme (second draft) to compensate for changes in inlet cooling water temperature.

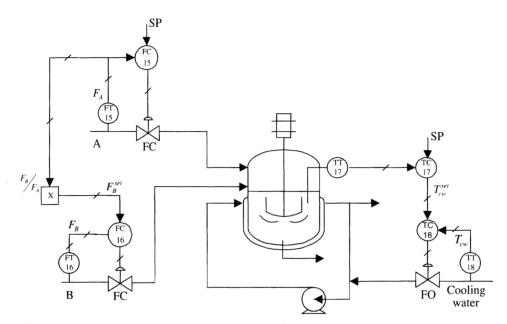

Figure 5-5.5   Proposed automatic control scheme (first draft) to compensate for changes in inlet cooling water temperature.

pares it to its set point, and decides on the set point to the slave controller. That is, the master controller decides on the inlet water temperature required, $T_{\mathrm{CW}}^{\mathrm{set}}$. Now suppose that the inlet water temperature is not equal to the set point, for example, $T_{\mathrm{CW}} > T_{\mathrm{CW}}^{\mathrm{set}}$. What would the slave controller do? Open the valve to add more water? Would this action make $T_{\mathrm{CW}} = T_{\mathrm{CW}}^{\mathrm{set}}$? The answer is, of course, no. The controller would open the valve, but T_{CW} would not change. Opening or closing the valve does not have any effect on T_{CW}. The controller would keep opening the valve until it winds up. This is a perfect example where the action taken by the controller does not affect its measurement. Remember M–D–A in Chapter 1? Remember we said that these three operations—measurement, decision, and action—must be in a loop? That is, the action (A) taken by the controller must affect its measurement (M). The scheme shown in Fig. 5-5.5 does not provide a closed loop, but rather, we have an open-loop.

Well, so this scheme does not work, but the idea is still valid; that is, learn as soon as possible that the cooling water temperature has changed. What about the scheme shown in Fig. 5-5.6? Go through the same analysis as previously and you will reach the same conclusion. That is, this last scheme still provides an open-loop. Opening or closing the valve does not affect the temperature where it is measured.

The earliest you can detect a change in cooling water and have a closed loop is any place in the recycle line or in the cooling jacket; Fig. 5-5.7 shows the transmitter installed in the recycle line, and Fig. 5-5.8 shows the transmitter installed in the jacket. Go through the previous analysis until you convince yourself that both of these schemes indeed provide a closed loop.

(c) For this part you again think of yourself as the controller. You know that as soon as the cooling system does not provide enough cooling, you must reduce the flow of reactants to the reactor. But how do you notice that you are short of cooling capacity? Certainly, if the temperature in the reactor or in the jacket reaches a high value, the cooling system is not providing the required cooling. But what is this value? Further analysis (thinking) indicates that the best indication of the cooling capacity is the opening of the cooling valve. When this valve is fully open,

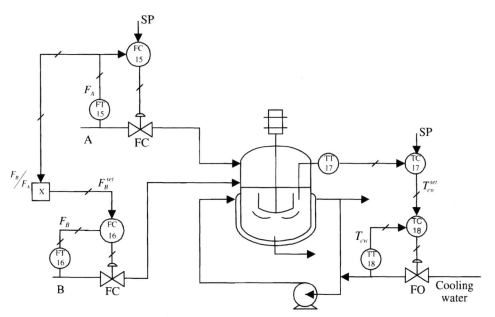

Figure 5-5.6   Proposed automatic control scheme (second draft) to compensate for changes in inlet cooling water temperature.

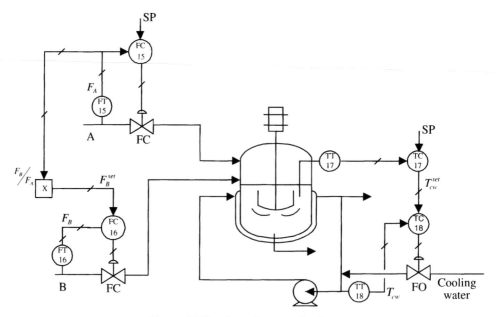

Figure 5-5.7   Cascade control scheme.

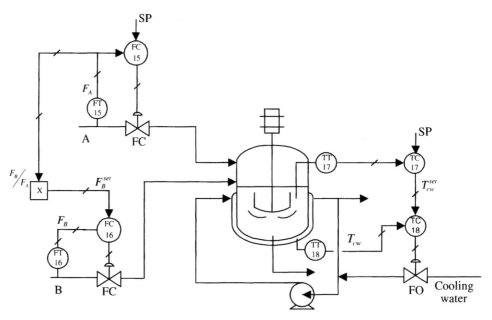

Figure 5-5.8   Cascade control scheme.

no more cooling is possible. At that time the temperature controller cannot do any more, and the process is out of control. Figure 5-5.9 shows what you may do as the controller.

The idea seems good but it is manual control, so how do I automate it now?

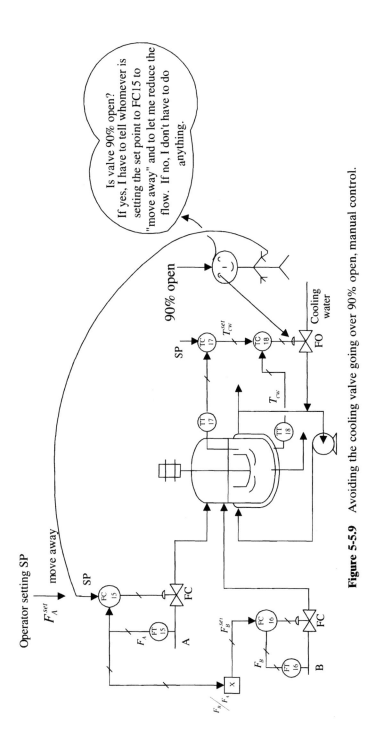

Figure 5-5.9 Avoiding the cooling valve going over 90% open, manual control.

101

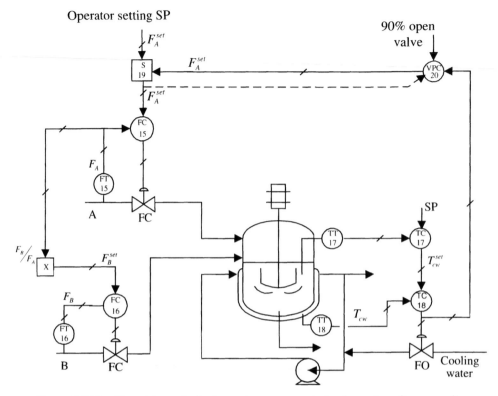

Figure 5-5.10   Override control scheme to compensate for loss of cooling capacity.

Figure 5-5.10 shows an override control scheme to do so. Let us explain. The output signal from TC18 that goes to the cooling valve is also used as the measurement to a controller, which we call VPC20. This controller compares the measurement, which indicates the valve position, with the set point and sends a signal to the flow controller, FC15. VPC20 is doing exactly what you were doing in Fig. 5-5.9. Note that before the signal from VPC20 gets to FC15, it goes through a selector. The selector is used to select which signal, the operator set or the one from VPC20, really goes to FC15. At this moment we do not know what type of selector (high or low) to use, so let's decide that now.

Under normal conditions the cooling valve is less that 90% open, say 65% open. As VPC20 "sees" that this valve is 65% open, it decides that the only way to make it open up to 90% is by asking for a lot of reactants and for that it increases the output signal, say up to 100%, to increase the set point to FC15. Obviously, under this condition there is plenty of cooling capacity left and there is no need to change the reactants flow required by the operator. The selector must be such that it selects the signal from the operator and not from VPC20. Since the signal from VPC20 is probably 100%, the selector must be a low selector; this is shown in Fig. 5-5.11. The selector is essentially used by VPC20 to tell the operator set point to "move away" and let it set the set point. Note the reset feedback indication from the selector to VPC20. Analyze this scheme until you fully understand the use of the selector.

Before leaving this example, there are a couple of additional things we need to

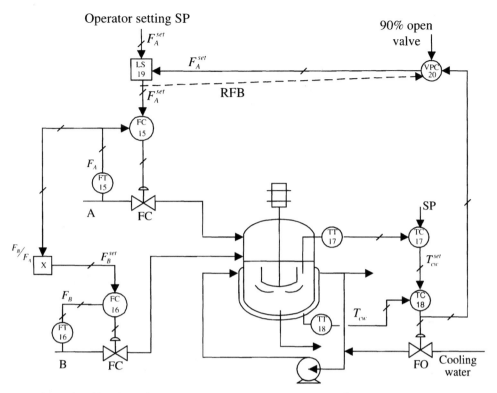

Figure 5-5.11   Override control scheme to compensate for loss of cooling capacity.

discuss. First, as you may recall from the discussion on cascade control in Chapter 4, FC15 must be set to remote set point. However, what would happen if the operator sets FC15 to local set point, or manual? That is, what would happen if VPC20 asks for a lower set point but FC15 is not in a remote set point? The answer is simply nothing, and the safety provided by VPC20 would not work; it is essentially not active. This is certainly not a safe operation. So what can we do? That is, how can we design a control system that allows the operator set FC15 in local set point or manual, and at the same time at any moment VPC20 would be able to reduce the flow of reactants? Think about it before reading further.

Figure 5-5.12 shows the new control scheme. In this scheme the output of VPC20 goes directly to the valve, not to the set point of FC15. So no matter what the mode of FC15, the decision to close the valve is after the controller. In this case the selector is also a low selector. The reader should convince himself or herself of this selection. If for safety considerations it is necessary to manipulate a flow, it is always good practice to go directly to the valve, not to the set point of a controller manipulating the valve. Note that the reset feedback goes to the two controllers feeding the selector.

Next, we need to address the set point of VPC20. The controller receives the output signal from TC18 to decide whether the valve is 90% open or not. Realize, however, that the cooling water valve is fail-opened (FO); therefore, the valve is 90% open when the signal is 10%. Thus the real set point in VPC20 must be 10%,

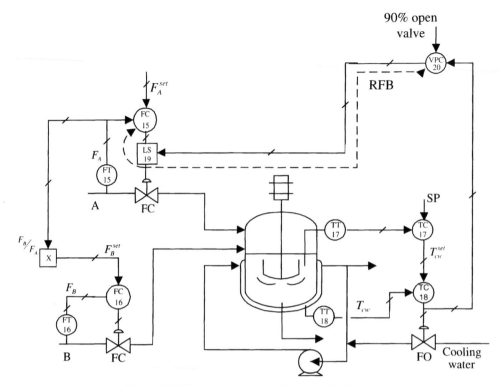

Figure 5-5.12   Another override control scheme.

as shown in Fig. 5-5.13. As an exercise, the reader may want to think about the action of VPC20.

Finally, as mentioned in Section 5-1, some DCSs allow the user to program the control scheme using software. Figure 5-5.14 shows a software program using the language presented in Section 5-1. The comments in each line help to relate the program to Fig. 5-5.13. For people with programming experience, it may be easier to design control schemes thinking first in programming terms.

Example 5-5.2.   Consider the reactor shown in Fig. 5-5.15, in which the irreversible and complete liquid reaction $A + B \rightarrow C$ occurs. Product C is the raw material for several process units downstream from the reactor. Reactant A is available from two sources. Because of a long-term contract, source 1 is less expensive than source 2. However, the contract is written with two limitations: a maximum instantaneous rate of 100 gpm and a maximum monthly consumption of 3.744×10^6 gal. If either of these limitations is exceeded, a very high penalty must be paid, and thus it is less expensive to use the excess from source 2. For example, if 120 gpm of reactant A is required, 100 gpm should come from source 1 and the other 20 gpm from source 2. Similarly, if on day 27 of the month 3.744×10^6 gal have been obtained from source 1, from then on, until the end of the month, all of reactant A should come from source 2. You may assume that the densities of each reactant, A and B, and of product C do not vary much and therefore can be assumed constant.

(a) Design a control system that will preferentially use reactant A from source

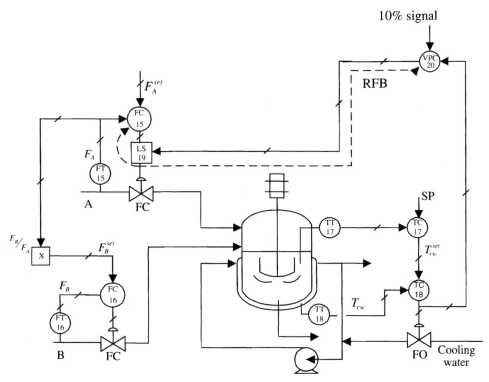

Figure 5-5.13   Override control scheme.

1   FA=AIN(1,FALO,FASPAN) ; reads in flow of stream A, FT15
2   FB=AIN(2,FBLO,FBSPAN) ; reads in flow of stream B, FT16
3   TR=AIN(3,TRLO,TRSPAN) ; reads in reactor's temperature, TT17
4   TJ=AIN(4,TJLO,TJSPAN) ; reads in jacket's temperature, TT18
5   FBSP=FA*RATIO ; calculates the SP for stream B, the RATIO is set by operator
6   VPB=PID(FB,FBSP,FBLO,FBSPAN) ; FC16
7   AOUT(1,VPB) ; outputs signal to valve in stream B
8   OUTTJSP=PID(TR,TRSP,TRLO,TRSPAN) ; master controller, TC17.
                                 ; set point TRSP is set by operator
9   TJSP=TJLO+OUTTJSP*TJSPAN/100 ; converts controller output of TC17 to
                                 ; engineering units of jacket temperature
10   VPCW=PID(TJ,TJSP,TJLO,TJSPAN) ; slave controller, TC18
11   AOUT(2,VPCW) ; outputs signal to cooling water valve
12   FAOVVP=PID(VPCW,10,0,100) ; VPC20
13   FAOPVP=PID(FA,FAOPSP,FALO,FASPAN) ; FC15, set point FAOPSP is set by
                                 ; operator
14   IF FAOPVP<FAOVVP THEN VPFA=FAOPVP ELSE VPFA=FAOVVP ; LS19
15   AOUT(3,VPFA) ; outputs signal to valve in stream A

Figure 5-5.14   Computer program of control scheme in Fig. 5-5.13.

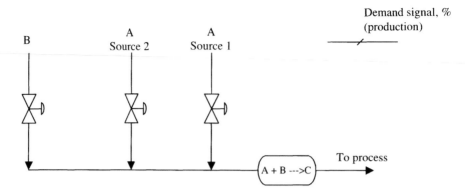

Figure 5-5.15   Reactor for Example 5-5.2.

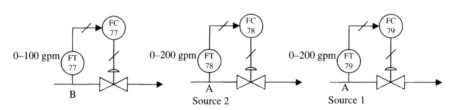

Figure 5-5.16   Flow loop installed in each stream.

1 and will not allow us to exceed contractual limitations. The feed ratio of A to B is 2:1 in gpm units.

(b) A few weeks after the control strategy designed in part (a) was put into operation, it was noticed that for some unknown reason, the supply pressure from source 2 was cut by the supplier every once in a while. Thus the flow controller manipulating the flow from source 2 would have to open the valve, and in some instances the valve would go wide open. At this moment there would not be enough flow from source 2 to satisfy the demand. It was decided that the correct action to take in this case, while the lawyers investigate—which may take a long time—is to obtain from source 1 whatever source 2 does not supply. Design a control strategy to accomplish this action. Be sure that your design is such that whenever source 2 provides the flow required, the scheme designed in part (a) is in effect.

SOLUTION: The first thing we'll do is to install a flow loop in each of the three streams; this is shown in Fig. 5-5.16 along with each transmitter's range. Let us work out this design by developing the software program first. Notice that the demand signal is the one that manipulates the flows; therefore, we first give this signal a significance. The significance can be anything that makes sense: for example, the flow of total A or the flow of B. Let's call it the flow of B; we then convert this signal from percent to units of flow of B:

```
1  FBSP = (demand signal) × 100/100 ; converts the demand
                                       signal from
                                     ; percent to units of
                                       flow stream B
```

We then send this set point to FC77:

```
2  FB = AIN(1,0,100) ; reads in the flow of B
3  VPB = PID(FB,FBSP,0,100) ; controls the flow of B (FC77)
4  AOUT(1,VPB) ; outputs signal to valve
```

Now that we know the flow of B, ratio the total flow of A, FA, to this flow:

```
5  FA = Ratio*FB ; the operator enters Ratio (FY75 in
   Fig. 5-5.17)
```

Before we set the set point for the flow of A from source 1, we must be sure that it does not exceed 100 gpm and that the totalized flow A from source 1 does not exceed 3.744×10^6 gal:

```
6  FA1 = AIN(2,0,200) ; reads in flow A from source 1
7  FA2 = AIN(3,0,200) ; reads in flow A from source 2
```

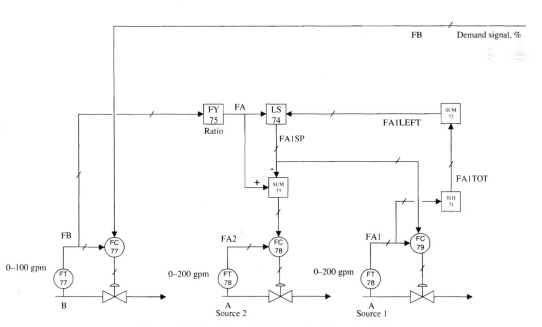

Figure 5-5.17 Control scheme for Example 5-5.2.

```
 8   IF FA < 100 THEN FA1SP = FA ELSE FA1SP = 100 ; FASP1 is
     ; set point to the controller in A source 1 (LS74)
 9   FA1TOT = TOT(FA1) ; TOT( ) is a function that totalizes
     ; its argument (TOT71)
10   FA1LEFT = 3.744 × 10⁶-FA1TOT ; calculates the flow from
     ; source 1 that is left to obtain during the month (SUM72)
11   IF FA1LEFT <= 0 THEN FA1LEFT = 0 ; checks to make sure
     ; that FA1LEFT < 0
12   IF FA1LEFT < FA1SP THEN FA1SP = FA1LEFT ; (LS74)
13   VPA1 = PID(FA1,FA1SP,0,200) ; FC79
14   AOUT(2,VPA1) ; outputs signal to valve in source 1
15   FA2SP = FA-FA1SP ; calculates SP for A source 2 (SUM73)
16   VPA2 = PID(FA2,FA2SP,0,200) ; FC78
17   AOUT(3,VPA2) ; outputs signal to valve in source 2
```

Lines 1 through 17 is the software program to accomplish the control objective. Figure 5-5.17 is the equivalent control scheme; it is wise to spend some time analyzing both ways.

Example 5-5.3. Consider the reactor shown in Fig. 5-5.18, where stream A reacts with water. Stream A can be measured but not manipulated. This stream is the by-product of another unit. The water enters the reactor in two different forms, as liquid and as steam. The steam is used to heat the reactor contents. It is necessary to maintain a certain ratio R between the total water and stream A into the reactor. It is also necessary to control the temperature in the reactor. It is important to maintain the ratio of total flow of water to flow of stream A below a value Y; otherwise, a very thick polymer may be produced, plugging the reactor.

A situation has occurred several times during extended periods of time in which the flow of stream A is reduced significantly. In this case the control scheme totally cuts the liquid water flow to the reactor to maintain the ratio. However, the steam flow to the reactor, to maintain temperature, still provides more water than required,

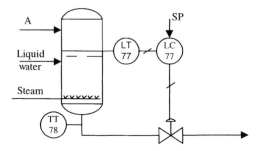

Figure 5-5.18   Reactor for Example 5-5.3.

and thus the actual ratio of water to stream A entering the reactor dangerously approaches Y. Design a control scheme to control the temperature in the reactor and another scheme to maintain the ratio of total water to stream A, while avoiding reaching the value of Y even if it means that the temperature deviates from the set point.

Figure 5-5.19 shows the temperature and ratio controls required. The temperature is controlled, manipulating the steam flow using a cascade control scheme. The flow of stream A is measured and multiplied by R, in MUL76, to obtain the total water flow required, F_{TW}. The steam flow is then subtracted from the total water to calculate the flow of liquid water required which it is then used as the set point to the liquid water controller.

Figure 5-5.20 shows a control scheme, added to the previous one, to avoid the ratio of total water to stream A exceeding the value of Y. In this scheme the actual flows of liquid water and steam are added, in SUM74, to obtain the water flow into the reactor (this total water should be the same as the output from MUL76 under normal conditions). The ratio of total water to stream A is calculated using DIV73. This ratio is then sent to a controller, RC95, with a set point set to Y, or to somewhat less than Y for safety. The outputs of RC95 and TC100 are compared in

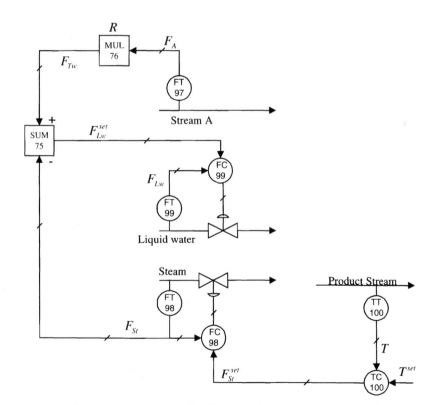

Figure 5-5.19   Temperature and ratio control for reactor of Example 5-5.3.

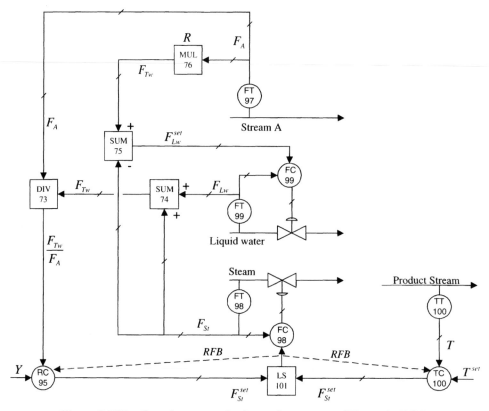

Figure 5-5.20 Complete control scheme for reactor of Example 5-5.3.

LS101 and the lowest selected as the set point to FC98. Under normal conditions TC100 will be selected. Only when the ratio of total water to stream A is above the set point to RC95 will the ratio controller reduce its output enough, in an effort to cut the steam, and thus, be selected. Note the reset feedback signals to RC95 and TC100.

Notice that the ratio of total water to stream A will be at, or close to, Y only after the liquid water flow has been reduced to zero; that is, the only water entering is the steam. Using this fact, Fig. 5-5.21 shows a simpler control scheme. In this case, F_A is multiplied by Y to obtain the maximum flow of water that could be fed, $F_{TW\,max}$. This scheme is simpler because there is no need to tune a controller. The reader may want to write the software program to implement the scheme shown in Fig. 5-5.21.

5-6 SUMMARY

In this chapter we have introduced the computation tools provided by manufacturers. An explanation for the need for scaling was given. A brief discussion of the significance, and importance, of field signals was also presented. We also presented the concepts, and applications, of ratio control, override control, and

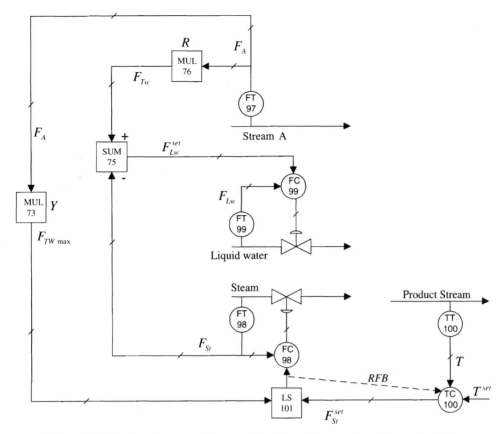

Figure 5-5.21   Another complete control scheme for reactor of Example 5-5.3.

selective control. These techniques provide a realistic and simple method for improving process safety, product quality, and process operation. Finally, the chapter concluded with three examples, to provide some hints on the design of control schemes.

REFERENCES

1. C. A. Smith and A. B. Corripio, *Principles and Practice of Automatic Process Control*, 2nd ed., Wiley, New York, 1997.

2. J. E. O'Meara, Oxygen trim for combustion control, *Instrumentation Technology*, March 1979.

3. T. J. Scheib and T. D. Russell, Instrumentation cuts boiler fuel costs, *Instrumentation and Control Systems*, November 1981.

4. P. Congdon, Control alternatives for industrial boilers, *InTech*, December 1981.

5. J. V. Becker and R. Hill, Fundamentals of interlock systems, *Chemical Engineering*, October 15, 1979.

6. J. V. Becker, Designing safe interlock systems, *Chemical Engineering*, October 15, 1979.

PROBLEMS

5-1. Consider the system shown in Fig. P5-1 to dilute a 50% by mass NaOH solution to a 30% by mass solution. The NaOH valve is manipulated by a controller not shown in the diagram. Because the flow of the 50% NaOH solution can vary frequently, it is desired to design a ratio control scheme to manipulate the flow of H_2O to maintain the dilution required. The nominal flow of the 50% NaOH solution is $200 \, lb_m/hr$. The flow element used for both streams is such that the output signal from the transmitters is related linearly to the mass flow. The transmitter in the 50% NaOH stream has a range of 0 to $400 \, lb_m/hr$, and the transmitter in the water stream has a range of 0 to $200 \, lb_m/hr$. Specify the computing blocks required to implement the ratio control scheme.

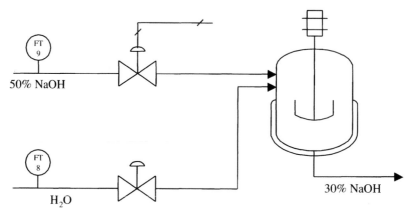

50% NaOH

H_2O

30% NaOH

Figure P5-1   Mixing process for Problem P5-1.

5-2. Consider the reactor shown in Fig. P5-2. This reactor is similar to a furnace in that the energy required for the reaction is provided by the combustion of a fuel with air (to simplify the diagram, the temperature control is not completely shown). Methane and steam are reacted to produce hydrogen by the reaction

$$CH_4 + 2H_2O \rightarrow CO_2 + 4H_2$$

The reaction occurs in tubes inside the furnace. The tubes are filled with a catalyst needed for the reaction. It is important that the reactant mixture be always steam-rich to avoid coking the catalyst. If enough carbon deposits over the catalyst, it poisons the catalyst. This situation can be avoided by ensuring that the entering mixture is always rich in steam. However, too much steam is also costly, in that it requires more energy (fuel and air) consumption. The engineering department has estimated that the optimum ratio R_1 (methane to steam) must be maintained. Design a control scheme which ensures that the required ratio be maintained and that during production rate changes, when it increases or decreases, the reactant mixture be steam-rich. Note that there is a signal that sets the methane flow required.

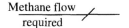

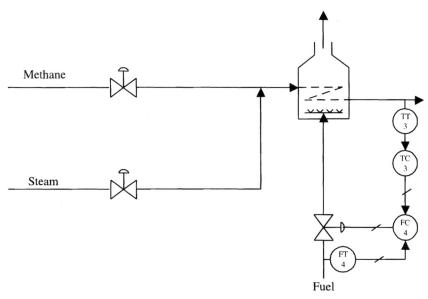

Figure P5-2 Reactor for Problem P5-2.

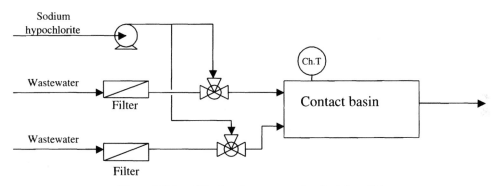

Figure P5-3 Chlorination process for Problem 5-3.

5-3. Chlorination is used for disinfecting the final effluent of a wastewater treatment plant. The Environmental Protection Agency (EPA) requires that
certain chlorine residual be maintained. To meet this requirement, the free
chlorine residual is measured at the beginning of the chlorine contact basin,
as shown in Fig. P5-3. An aqueous solution of sodium hypochlorite is added
to the filter effluent to maintain the free chlorine residual at the contact basin.
The amount of sodium hypochlorite required is directly proportional to the
flow rate of the effluent. The wastewater plant has two parallel filter effluent
streams, which are combined in the chlorine contact basin. Sodium hypochlorite is added to each stream based on free chlorine residual in the basin.

(a) Design a control scheme to control the chlorine residual at the beginning of the basin.

(b) Due to a number of reactions occurring in the contact basin, the chlorine residual exiting the basin is not equal to the chlorine residual entering the basin (the one being measured). It happens that the EPA is interested in the exiting chlorine residual. Thus, a second analyzer is added at the effluent of the contact basin. Design a control scheme to control the effluent chlorine residual.

5-4. Consider the tank shown in Fig. P5-4. In this tank three components are mixed in a given proportion so as to form a stock that will be supplied to another process. For a particular formulation the final mixture contains 50 mass % of A, 30 mass % of B, and 20 mass % of C. Depending on its demand, the other process provides the signal to the pump. Design a control system to control the level in the tank and at the same time maintain the correct formulation.

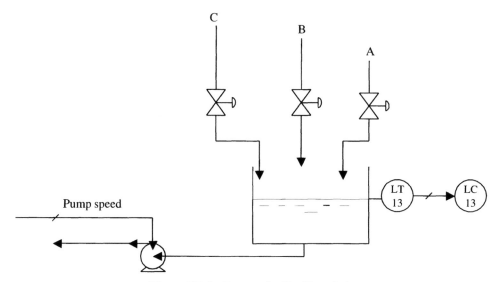

Figure P5-4   Process for Problem 5-4.

5-5. Fuel cells are used in spacecraft and proposed extraterrestrial bases for generating power and heat. The cell produces electric power by the reaction between liquid hydrogen and liquid oxygen:

$$2H_2 + O_2 \rightarrow 2H_2O$$

Design a ratio controller to maintain the flows of liquid hydrogen and oxygen into the cell in the exact stoichiometric ratio (both hydrogen and oxygen are valuable in space, so we cannot supply either in excess). Calculate the design flows of hydrogen and oxygen required to produce 0.5 kg/h of water, and the design ratio of oxygen to hydrogen flow. Sketch a ratio control scheme that will manipulate the flow of oxygen to maintain the exact stoichiometric ratio between the two flows. You may assume that the signals from the flow transmitters are linear with the mass flow rates. Calculate reasonable ranges for the flow transmitters and the ratio in terms of the transmitter signals.

5-6. Consider a furnace, shown in Fig. P5-6, consisting of two sections with one common stack. In each section the cracking reaction of hydrocarbons with steam takes place. Manipulating the fuel to the particular section controls the temperature of the products in each section. Manipulating the speed of a fan installed in the stack controls the pressure in the stack. This fan induces the flow of flue gases out of the stack. As the pressure in the stack increases, the pressure controller speeds up the fan to lower the pressure.

(a) Design a control scheme to ratio the steam flow to the hydrocarbons flow in each section. The operating personnel is to set the hydrocarbons flow.

(b) During the last few weeks the production personnel have noticed that the pressure controller's output is consistently reaching 100%. This indicates that the controller is doing the most possible to maintain pressure control. However, this is not a desirable condition since it means that the pressure is really out of control—not a safe condition. A control strategy must be designed such that when the pressure controller's output is greater than 90%, the flow of hydrocarbons starts to be reduced to maintain the output at 90%. As the flow of hydrocarbons is reduced, less fuel is required to maintain exit temperature. This, in turn, reduces the pressure in the stack and the pressure controller will slow down the fan. Whenever the controller's output is less than 90%, the feed of hydrocarbons can be whatever the operating personnel require.

It is known that the left section of the furnace is less efficient than the right section. Therefore, the correct strategy to reduce the flow of hydrocarbons calls for reducing the flow to the left section first, up to 35% of the flow set by operating personnel. If further reduction is necessary, the flow of hydrocarbon to the right section is then reduced, also up to 35%

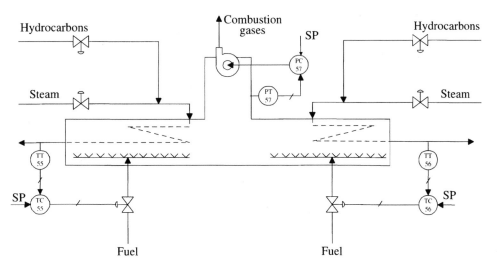

Figure P5-6 Furnace for Problem 5-6.

of the flow set by operating personnel. (If even further reduction is necessary, an interlock system would then drop the furnace offline.) Design a control strategy to maintain the pressure controller's output below 90%.

5-7. Consider the process shown in Fig. P5-7. Mud is brought into a storage tank, T3, from where it is pumped to two filters. Manipulating the exit flow controls the level in the tank. This flow must be split between the two filters in the following known ratio:

$$R = \frac{\text{flow to filter 1}}{\text{total flow}}$$

The two flow transmitters and control valves shown in the figure cannot be moved from their present locations, and no other transmitters or valves can be added. Design a control system that controls the level in T3 while maintaining the desired flow split between the two filters.

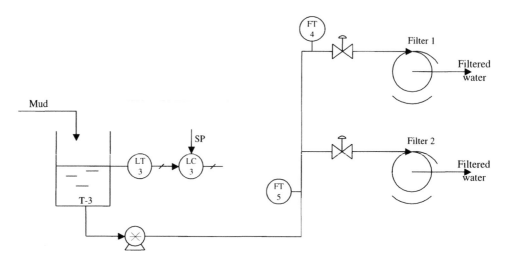

Figure P5-7   Process for Problem 5-7.

5-8. Consider the furnace of Fig. P5-8, where two different fuels, a waste gas and fuel oil, are manipulated to control the outlet temperature of a process fluid. The waste gas is free to the operation, and thus it must be used to full capacity. However, environmental regulations dictate that the maximum waste gas flow be limited to one-fourth of the fuel oil flow. The heating value of the waste gas is HV_{wg}, and that of the fuel oil is HV_{oil}. The air/waste gas ratio is R_{wg} and the air/fuel oil ratio is R_{oil}.

(a) Design a cross-limiting control scheme to control the furnace product temperature.

(b) Assume now that the heating value of the waste gas varies significantly as its composition varies. It is difficult to measure on-line the heating value of this gas; however, laboratory analysis has shown that there is definitely a correlation between the density of the gas and its heating value.

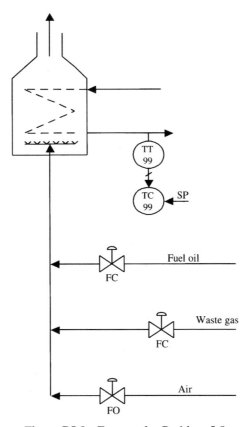

Figure P5-8   Furnace for Problem 5-8.

There is a densitometer available to measure the density, and therefore the heating value is known. Adjust the control scheme design in part (a) to consider variations in HV_{wg}.

(c) For safety reasons it is necessary to design a control scheme such that in case of loss of burner flame, the waste gas and fuel oil flows cease; the air dampers must open wide. Available for this job is a burner switch whose output is 20 mA as long as the flame is present and whose output drops to 4 mA as soon as the flame stops. Design this control scheme into the preceding one.

5-9. Consider the process shown in Fig. P5-9. In this process a liquid product is separated from a gas; the gas is then compressed. Drum D103 provides the necessary residence time for the separation. The pressure in the drum is controlled at 5 psig, as shown in the figure. Another pressure controller opens the valve to the flare if the drum pressure reaches 8 psig. There is always a small amount of recycle gas to the drum. The turbine driving the compressor is rather old, and for safety considerations its speed must not exceed 5600 rpm or drop below 3100 rpm. Design a control scheme that provides this limitation.

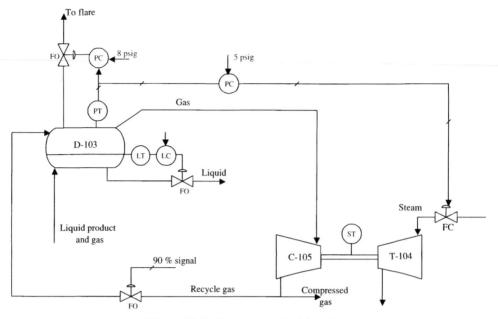

Figure P5-9   Process for Problem 5-9.

5-10. Consider the process shown in Fig. P5-10. The feed to the reactor is a gas and the reactor produces a polymer. The outlet flow from the reactor is manipulated to control pressure in the reactor. Exiting the reactor is polymer with entrained gas. This outlet flow goes to a separator, which provides enough residence time to separate the gas from the polymer. The polymer product is manipulated to control the level in the separator; the gases flow out of the separator freely. These gases contain the unreacted reactants and an amount of wax components that have been produced. The gases are compressed before returned to the reactor. A portion of the gases are cooled and mixed with the reactor effluent to control the temperature in the separator, as shown in the figure. If the temperature in the separator is too high, the wax components will exit with the gases. This wax will damage the compressor and it is why cyclones are installed in the recycle line. If the temperature in the separator is too low, the polymer will not flow out of the separator. Thus, the separator temperature must be controlled.

When the separator temperature increases, the temperature controller opens the recycle valve to increase the flow of cool gas. Under some significant upsets, as when a new polymer product is being produced, the recycle valve may go wide open in an effort to control the temperature. At this time the operator manually opens the chilled water valve to the gas coolers. This action reduces the gas temperature, providing more cooling capacity to the separator and thus the gas valve can close. Design a control scheme that provides this operation automatically.

5-11. Figure P5-11 shows a system to filter an oil before processing. The oil enters a header in which the pressure is controlled, for safe operation, by manipu-

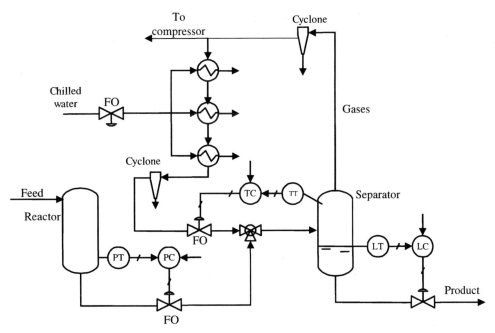

Figure P5-10   Process for Problem 5-10.

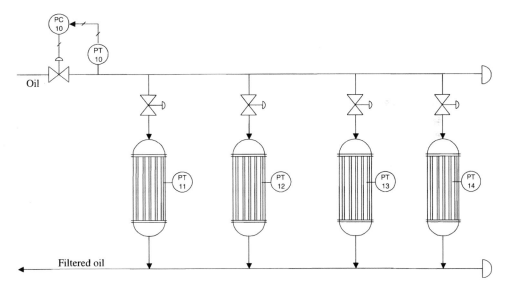

Figure P5-11   Filters for Problem 5-11.

lating the inlet valve. From the header, the oil is distributed to four filters. The filters consist of a shell with tubes inside, similar to heat exchangers. The tube wall is the filter medium through which the oil must flow to be filtered. The oil enters the shell and flows through the medium into the tubes. As time passes, the filter starts to build up a cake, and consequently, the oil pressure required for flow increases. If the pressure increases too much, the walls may collapse. Thus, at some point the filter must be taken out and cleaned. Under normal conditions, three filters can handle the total oil flow.

(a) Design a control system to maintain the total oil flow through the system.

(b) Design a control system so that as the oil pressure in the shell side of each filter increases above a predetermined value, the oil flow to that filter starts to decrease. Once the feed valve to the filter is 10% open, an interlock system will shut-down the filter for cleaning. The total oil flow through the system must still be maintained.

5-12. Figure P5-12 shows a process often found in chemical plants. R101 is a reactor where a high-pressure gas is generated. It is necessary to transfer the gas to a low-pressure vessel, V102, at about 50 psig. As an energy-saving measure, the gas pressure is dropped across a power recovery turbine, T102. The work produced in T102 is used to drive a compressor, C102. However, the work produced in T102 is usually not enough to run C102; therefore, a steam turbine, T103, is connected in series with T102 to provide the necessary work. The figure shows the control systems to control the pressure in R101, and the pressure of the gas leaving the compressor. ST16 is a speed transmitter connected to the turbine's shaft, SC16 is a controller controlling the shaft's rotational speed. Out of R101 there is a line that bypasses T102 and goes directly to V102. This line is used in case T102 is down, or any emergency develops and

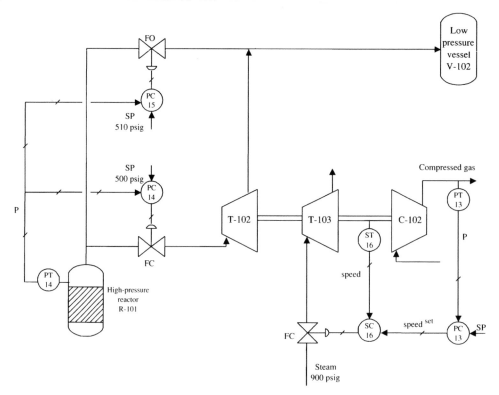

Figure P5-12   Process for Problem 5-12.

it is necessary to quickly relief the pressure in T102. The set point to PC14 is 500 psig, while the one to PC15 is 510 psig. A condition occurs when the gas produced in R101 increases significantly thus, increasing the pressure. In this case PC14 opens more the valve to T102 to relief the pressure. When this occurs, the steam valve to T103 backs off to control the pressure of the compressed gas. It has been determined that if the steam valve is less than 10% open severe mechanical damage can occur in T103. Design a control scheme to avoid this condition while still controlling the pressure of the compressed gas.

5-13. In Example 5-3.2 the constraint control scheme shown in Fig. 5-2.4 was presented and discussed. In the discussion of the scheme it was mentioned: "If at any time the operating personnel were to set the flow controller FC11 in local set point or in the manual mode (i.e., off remote set point), the safety provided by TC13 and PC14 would not be in effect. This would result in an unsafe and unacceptable operating condition." Modify the control scheme shown such that even when FC11 is off cascade, or remote set point, controllers TC13 and PC14 still provide the necessary safety override.

5-14. Consider the turbine/compressor process shown in Fig. P5-13. The motive force for the turbine, T30, is a high-pressure gas, and the compressor, C30, compresses a refrigerant gas. The operator sets the valve position of the high-pressure gas valve, which in turn results in a certain compressor speed. A lag unit is used to avoid sudden changes in the high-pressure valve position. Under normal operating conditions the valve should respond to the operator's set value. However, there are some special conditions that a control system must guard against.

1. Under normal conditions the pressure in the refrigerant gas line is about 15 psig. However, during startup and other circumstances, the pressure in

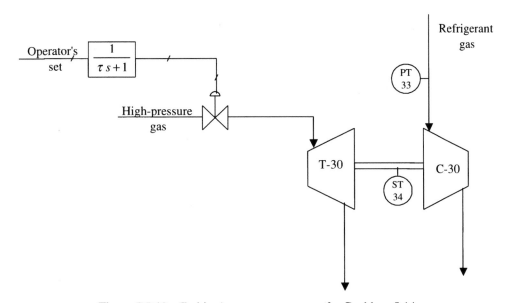

Figure P5-13   Turbine/compressor process for Problem 5-14.

the line tends to drop below 8 psig, which is a dangerous condition to the compressor. In this case the compressor velocity must be reduced to pull-in less gas, thus increasing the pressure in the line. The lowest safe pressure in the refrigerant line is 8 psig.

2. Due to mechanical difficulties, the compressor velocity must not exceed 95% of its maximum rated velocity. Also, it must not drop below 50% of its maximum velocity.

Design a control scheme that avoids violating constraints 1 and 2.

5-15. Consider the compressor shown in Fig. P5-14. This two-stage compressor has two different suction points. In each suction line there is a volumetric flow meter calibrated at 0°C and 1 atm, a pressure transmitter, and a temperature transmitter. An important consideration in the control of the compressor is to avoid the surge condition. Figure P5-14 also shows a curve indicating the minimum inlet flow, in acfm (actual ft³/min), required for a given inlet pressure, to avoid surge. Each stage can go into surge independently. Under normal operating conditions the operator sets the position to each suction valve. However, for safety considerations, the operator must not close the valves below the surge limit. Design a control scheme to avoid closing the valves below the surge limit. Also, for safety considerations, it is permissible to open the valves very fast. However, in closing the valves, it must be done slowly. Design this constraint into your previous design.

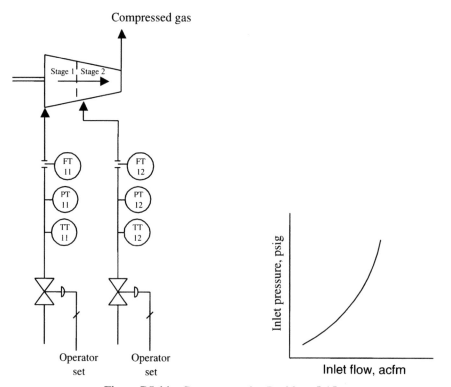

Figure P5-14 Compressor for Problem 5-15.

5-16. High-pressure water is often used to force petroleum crude out of oil wells. Figure P5-15 shows the process to prepare the water and its injection into the wells. The water must be filtered from suspended solids and deoxygenated before entering the wells. The figure shows three filters—F43, F44, and F45—where the suspended solids are removed. Two filters are enough to filter the water. As a filter plugs, due to the solids removed from the water, the pressure drop required to flow through it increases. Once this pressure drop reaches a critical value, the filter must be removed from service and cleaned. The cleaning, which consists of back-flushing with a solvent, is a relatively fast operation. From the filters the water flows to a scrubber, T46, where the dissolved oxygen is stripped. The scrubber consists of a top section in which the water flows countercurrent to natural gas, which is the stripping agent. The internals of this top section are similar to that of a plate distillation column. From the top section the water flows to the bottom section, which looks like a surge tank. The level in the tank must be controlled. The dissolved O_2 in the water out of this tank must also be controlled. From the scrubber the water is then pumped to three oil wells.

(a) Design a control scheme to control the level of water in the bottom section of the tank.

(b) Design an override control scheme that reduces the water flow through any filter to avoid the pressure drop reaching the critical value. Once the water valve in the filter reaches a minimum prescribed opening, sequential logic takes over to stop the operation and start the cleaning cycle. You are not required to show this.

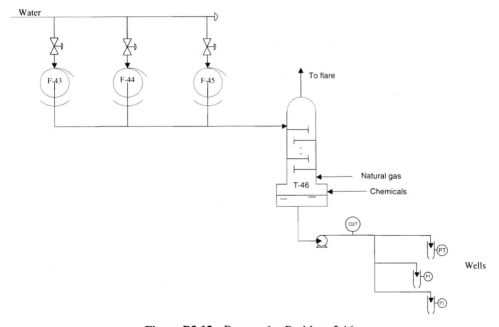

Figure P5-15 Process for Problem 5-16.

(c) Design a control scheme to control the dissolved O_2 in the water leaving the scrubber. Most of the control action can be obtained by manipulating the natural gas; however, sometimes this gas by itself is not enough to reach the set point. In such cases chemicals are used to reduce the O_2 to its desired value. The natural gas is less expensive than the chemicals. It is known that the most important disturbance to this control is the water flow to the scrubber.

(d) Design a control scheme to control the flow of water to each well. An important consideration is the water pressure in each well. As the well ages, or internal disturbances occur, the pressure in the well increases. If the pressure reaches a certain value, it may crack the well. Thus the pressure in each well must be considered in this control scheme.

5-17. Consider the furnace shown in Fig. P5-16. This furnace is used to partially vaporize water. Liquid water enters the furnace and a mixture of saturated liquid and vapor exits. The mixture then goes into a tank where residence time is provided to separate the saturated vapor from the saturated liquid. The process engineers talk about the efficiency of this process. They define *efficiency* as the fraction of the liquid that is vaporized; the desired efficiency is 80%. Design a control scheme to maintain the efficiency at this desired value.

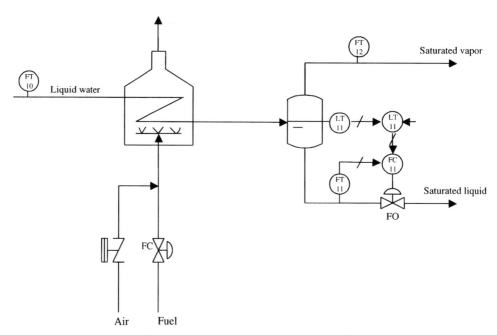

Figure P5-16   Process for Problem 5-17.

5-18. Consider the bottom of the distillation column shown in Fig. P5-17. There are two streams leaving the column. One of the streams is manipulated by user 1 to satisfy its own needs. The other stream is manipulated to maintain the liquid level in the bottom of the column. Under some upset conditions the level drops enough so that the level controller closes the valve; when this happens,

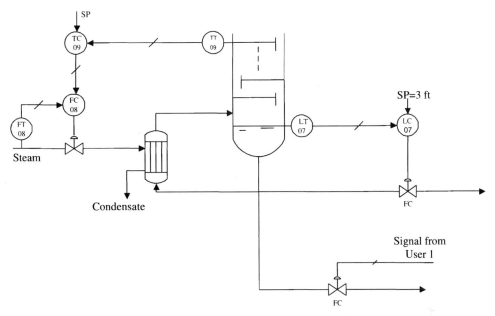

Figure P5-17   Process for Problem 5-18.

level control is lost. If the level ever drops below 1.5 ft, it would be very difficult to have enough flow through the reboiler to provide boil-up to the
column; this condition would be catastrophic to operation of the column. The
level transmitter is calibrated for 1 to 5 ft. What would you propose to avoid
this condition? Design the control scheme to implement your proposal.

5-19. Consider the three-stage compressor shown in Fig. P5-18. The figure shows
the control schemes associated with the compressor. PC25 controls the discharge pressure from stage 2 (set point = 65 bar), and PC26 controls the discharge pressure from stage 3 (set point = 110 bar). PV28 opens to discharge
the gas to the flare stack if the suction pressure rises above 21 bar.

There are several interlock systems that drop the compressor offline for
various reasons: (1) low suction pressure to stage 1 (<14 bar), (2) high-
pressure rise across stage 2 (>40 bar), and (3) high-pressure rise across stage
3 (>65 bar). The compressor drops offline often, and anytime it happens it
takes approximately 3 hours before returning to normal operation. This
amount of time represents a significant loss to the operation. Obviously,
sending gas to the flare is another loss. Design the necessary override control
scheme to minimize these losses. Specify the necessary set points.

Steady-state information:

P suction stage 1 = 18 bar	P discharge stage 1 = 42 bar
P discharge stage 2 = 65 bar	P discharge stage 3 = 110 bar
P rise across stage 2 = 23 bar	P rise across stage 3 = 45 bar

5-20. Consider the process shown in Fig. P5-19. This process is used to manufacture
product E from the reaction of A and B. The output from the reactor is

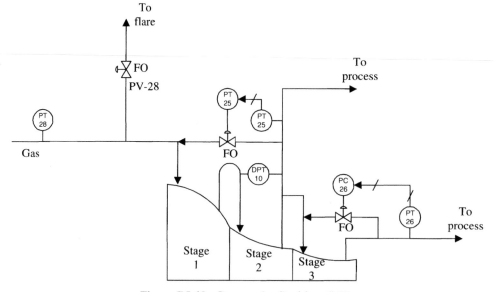

Figure P5-18 Process for Problem 5-19.

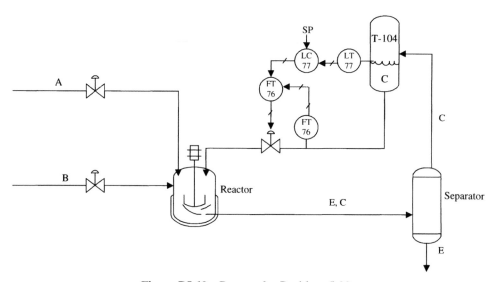

Figure P5-19 Process for Problem 5-20.

product E and some unreactants, mainly A, referred to as liquid C. E and C are separated and liquid C is recycled back to tank T104 to be fed back to the reactor. The amount of B fed to the reactor depends on the amount of A and on the amount of C fed to the reactor. That is, there must be some B to react with the A fed, given by the ratio $R_1 = B/A$, and some B to react with the C fed, given by the ratio $R_2 = B/C$. You may assume that all the flowmeters provide a signal related to mass flow. Design a control scheme to control the total flow T (lb/min) into the reactor.

CHAPTER 6

BLOCK DIAGRAMS AND STABILITY

This chapter presents a discussion of block diagrams and control loop stability. It is important to present the development of block diagrams because they are used in the study of stability, in the design of feedforward controllers (Chapter 7), in understanding the Smith predictor dead-time compensation (Chapter 8), and in understanding multivariable control (Chapter 9). The presentation of stability is done minimizing the mathematics and emphasizing the physical significance.

6-1 BLOCK DIAGRAMS

Block diagrams show graphically how the process units and the instrumentation interact to provide closed-loop control. These diagrams are composed of three elements.

1. *Arrows:* ⟶

   The arrows indicate either variables or signals.

2. *Blocks:* I ⟶ G ⟶ O

   Every block has an input, I, and an output, O. Inside the box we write the equation that describes how the input affects the output. In control work, this equation is a transfer function. Remember: *The transfer function tells us how the input affects the output*; that is,

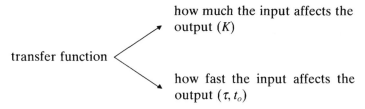

transfer function — how much the input affects the output (K)

— how fast the input affects the output (τ, t_o)

3. *Circles:*   A $\xrightarrow{\ +\ }$ ◯ $\xrightarrow{\ C\ }$

|
B

Circles have at least two inputs. They represent the algebraic summation of the inputs, or C = A − B.

Let us now look at the development of block diagrams for two processes.

Example 6-1.1. Consider the heat exchanger control system shown in Fig. 6-1.1. The author starts the block diagram by writing an arrow (Fig. 6-1.2) that represents the controlled variable in engineering units, in this case the outlet temperature T. Note that not only is the variable T written, but also, and very important, the units are written. Recall that in Chapter 5 we always indicated the significance of every signal in a control diagram; the same should (must?) be done in block diagrams.

Once the arrow is drawn, we continue by drawing the feedback loop until we return to the variable just drawn. The first device of the loop is the sensor/transmitter (Fig. 6-1.3). Generally, the letter G is used to indicate transfer functions; however, by convention, the letter H is used to represent the transfer function of transmitters. Note that the letter c is used to represent the output signal from the transmitter. This is to remind you that for the controller, this is the real "controlled

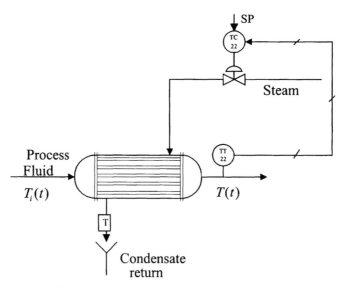

Figure 6-1.1   Heat exchanger control system.

$$\xrightarrow{\quad T,\ ^{\circ}F \quad}$$

Figure 6-1.2   Arrow representing the controlled variable in engineering units.

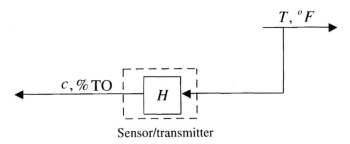

Figure 6-1.3   Block diagram showing sensor/transmitter.

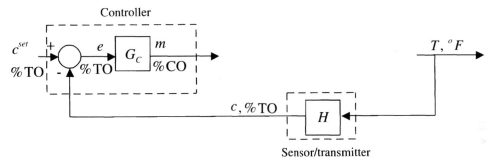

Figure 6-1.4   Block diagram with controller added.

variable," as discussed in Chapters 2 and 3. We use the generic %TO as the units of c.

We now continue moving around the feedback loop by drawing the controller (Fig. 6-1.4). The controller is by convention drawn using a circle and a block. The circle indicates that the first thing the controller does is to subtract the measurement, c, from the set point, c^{set}, to look for an error, e; the controller equation then acts on the error. G_c is the transfer function of the controller, given by Eq. (3-2.5), (3-2.11), or (3-2.13), depending on the type of controller. Note that the letter m is used to indicate the controller output and to remind you that for the controller, this is the real "manipulated variable."

From the controller we move to the final control element, a valve in this case (Fig. 6-1.5). G_V is the transfer function that describes how the controller output, in %CO, affects the steam flow, in lb/min.

Finally, we move to the process unit, the heat exchanger (Fig. 6-1.6). G_1 is the transfer function that describes how the steam flow, in lb/min, affects the outlet temperature T. The figure shows the "closed" loop.

The block diagram of Fig. 6-1.6 shows only one input, the set point c^{set}. However, we know that there are other inputs, the disturbances, that also affect the controlled variable. The disturbances are drawn as shown in Fig. 6-1.7. This figure shows three disturbances: the inlet temperature T_i, the process flow F_p, and the upstream pressure from the valve, P_u. G_2 is the transfer function that describes how T_i affects T. G_3 is the transfer function that describes how F_p affects T. Finally, G_4 is the

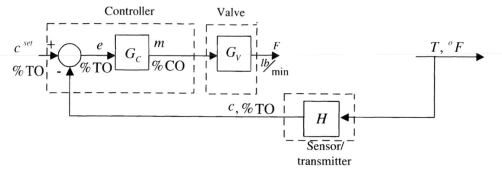

Figure 6-1.5 Block diagram with valve added.

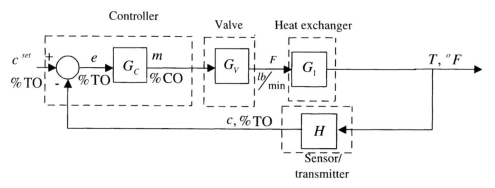

Figure 6-1.6 Block diagram showing control loop.

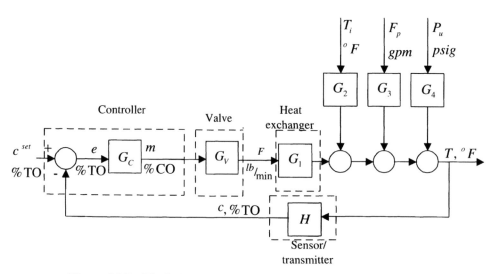

Figure 6-1.7 Block diagram showing control loop and disturbances.

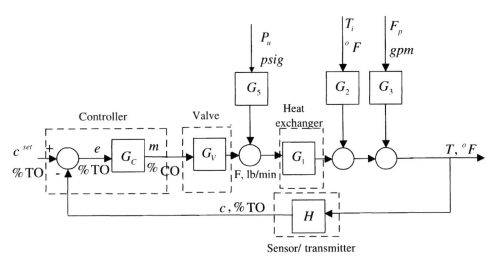

Figure 6-1.8 Block diagram showing control loop and disturbances.

transfer function that describes how P_u affects T. All three transfer functions, as well as G_1, describe the heat exchanger.

A detail analysis of the process shows that when P_u changes, it first affects the steam flow F and then affects T. So the following question develops: How do we present these effects in the block diagram? Figure 6-1.8 shows these effects. G_5 is the transfer function that describes how P_u affects F. The figure shows that P_u first affects F and then F affects T. Comparing Figs. 6-1.7 and 6-1.8, we see that $G_4 = G_5 G_1$. Thus we can draw block diagrams in different ways as long as they make physical sense. Look at Fig. 6-1.9; can we draw the block diagram that way? Yes or no? Why?

Before proceeding to another example it is important to show some simplifications of the block diagram just drawn. Remember that the transfer functions are in terms of Laplace transforms. The reason for using these transforms is that we can work with them using algebra instead of using differential equations. Thus we can use the rules of algebra with the block diagrams. The three diagrams shown in Fig. 6-1.10 can be developed starting from Fig. 6-1.7. In the figure

$G_M = G_V G_1 H$; transfer function that describes how the controller output m affects the transmitter output c

$G_{D1} = G_2 H$; transfer function that describes how the inlet temperature T_i affects the transmitter output c

$G_{D2} = G_3 H$; transfer function that describes how the process flow F_P affects the transmitter output c

$G_{D3} = G_4 H$; transfer function that describes how the upstream pressure from the valve P_U affects the transmitter output c

Example 6-1.2. Consider the control system for the drier shown in Fig. 6-1.11, which dries rock pellets. The rock is obtained from the mines, crushed into pellet size, and washed in a water-intensive process. These pellets must be dried before

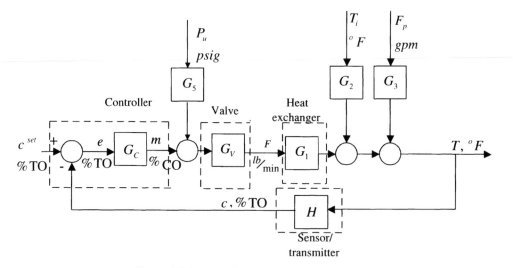

Figure 6-1.9 Another way to draw Fig. 6-1.8.

feeding them into a reactor. The moisture of the exiting pellets must be controlled. The moisture is measured and a controller manipulates the speed of the conveyor belt to maintain the moisture at its set point. Let us draw the block diagram for this control scheme.

As shown in Example 6-1.1, we start by drawing an arrow depicting the controlled variable in engineering units, Mois (%) (Fig. 6-1.12a). We then continue "around the loop" by adding the sensor/transmitter (Fig. 6-1.12b), then the controller (Fig. 6-1.12c), then the final control element, which in this case is a conveyor belt (Fig. 6-1.12d), and finally, the process unit (Fig. 6-1.12e). This completes the block diagram of the control loop itself. Figure 6-1.12f shows the diagram when two disturbances, the heating value of the fuel, HV, and the inlet moisture of the pellets, IMois, are added. Figure 6-1.12f shows that the block diagram is very similar to that of Fig. 6-1.7. Algebraic simplification of Fig. 6-1.12 would yield a diagram similar to Fig. 6-1.10c.

6-2 CONTROL LOOP STABILITY

Once we have learned how to draw block diagrams, the subject of control loop stability can be addressed. We are particularly interested in learning the maximum gain that puts the process to oscillate with constant amplitude. In Chapter 3 we mentioned that this gain is called the the *ultimate gain*, K_{CU}. Above this gain the loop is unstable (you may even say that at this value the loop is already unstable); below this gain the loop is stable.

Let us consider the heat exchanger, shown in Fig. 6-1.1, and its block diagram, shown in Fig. 6-1.7. The transmitter is calibrated from 50 to 150°F. Suppose that the following are the transfer functions of each block in the "loop":

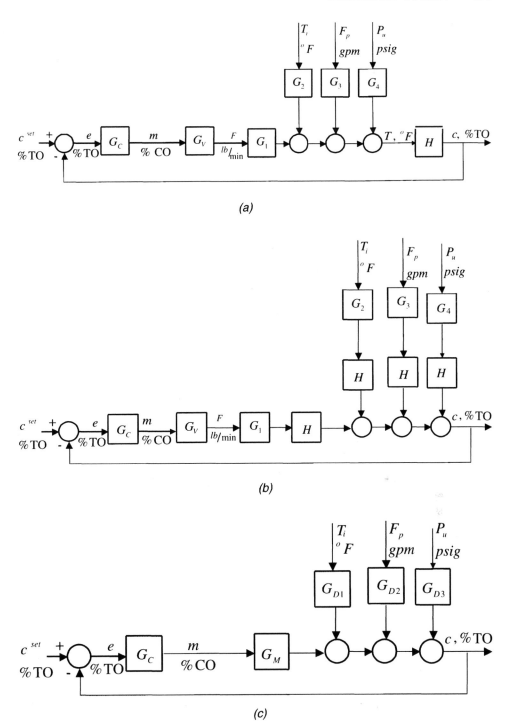

(a)

(b)

(c)

Figure 6-1.10   Algebraic simplification of Fig. 6-1.7.

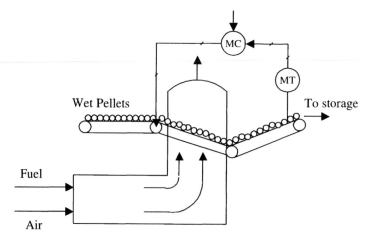

Figure 6-1.11 Phosphate pellets drier.

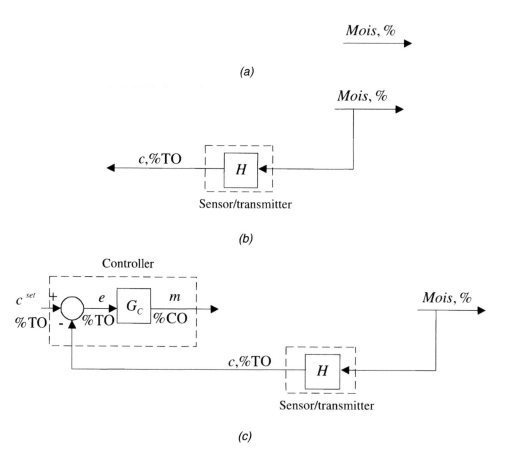

Figure 6-1.12 Developing the block diagram of the drier control system.

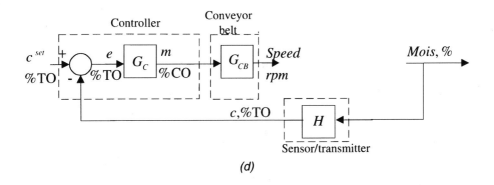

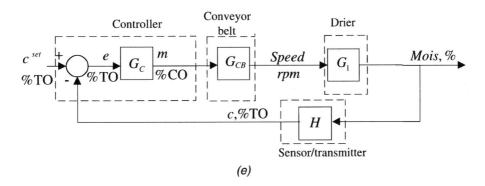

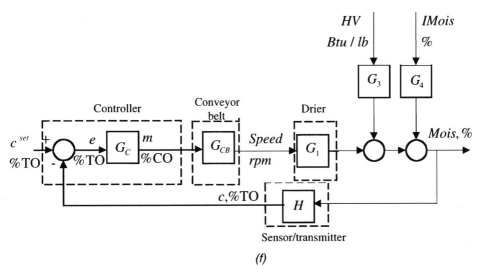

Figure 6-1.12   *Continued.*

$$G_V = \frac{0.016}{3s+1} \qquad G_1 = \frac{50}{30s+1} \qquad H = \frac{1.0}{10s+1}$$

The time constants are in seconds. The gain of 1.0 in H is obtained by

$$\frac{(100-0)\%\mathrm{TO}}{(150-50)°\mathrm{F}} = \frac{100\%\mathrm{TO}}{100°\mathrm{F}} = 1.0\frac{\%\mathrm{TO}}{°\mathrm{F}}$$

To study the stability of any control system, control theory says that we need only to look at the *characteristic equation* of the system. For block diagrams such as the one shown in Fig. 6-1.7, the characteristic equation is given by

$$1 + G_C G_V G_1 H = 0 \tag{6-2.1}$$

That is, the characteristic equation is given by one (1) plus the multiplication of all the transfer functions in the loop, all of that equal to zero (0). Thus

$$1 + \frac{(0.016)(50)(1)G_C}{(3s+1)(30s+1)(10s+1)} = 0 \tag{6-2.2}$$

or

$$900s^3 + 420s^2 + 43s + (1+0.8G_C) = 0 \tag{6-2.3}$$

Note that the transfer functions of the disturbances are not part of the characteristic equation, and therefore they do not affect the stability of the loop.

Let us first look at the stability when a P controller is used; for this controller $G_C = K_C$. The characteristic equation is then

$$900s^3 + 420s^2 + 43s + (1+0.8G_C) = 0 \tag{6-2.4}$$

This equation is a polynomial of third order; therefore, there are three roots in this polynomial. As we may remember, these roots can be either real, imaginary, or complex. Control theory and mathematics says that for any system to be stable, *the real part of all the roots must be negative*; Fig. 6-2.1 shows the stability region. Note from Eq. (6-2.4) that *the locations of the roots depend on the value of K_C, which is the same thing as saying that the stability of the loop depends on the tuning of the controller.*

If there were two roots on the imaginary axis (they come in pairs of complex conjugates) and all other roots were on the left side of the imaginary axis, the loop would be oscillating with a constant amplitude. The value of K_C that generates this case is K_{CU}.

There are several ways to proceed from Eq. (6-2.4), and control textbooks [1] are delighted to show you so. In this book we are interested only in the final answer, that is, K_{CU}, not in the mathematics. For this case, which we call the *base case*, the K_{CU} value and the period at which the loop oscillates, which in Chapter 2 we called the *ultimate period T_U* are

$$K_{CU} = 23.8\frac{\%\mathrm{CO}}{\%\mathrm{TO}} \quad \text{and} \quad T_U = 28.7\,\mathrm{sec}$$

Let us now learn what happens to these values of K_{CU} and T_U as terms in the loop change.

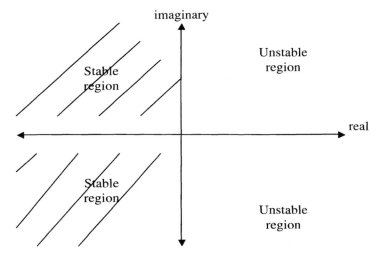

Figure 6-2.1 Roots of the characteristic equation.

6-2.1 Effect of Gains

Let us assume that a new transmitter is installed with a range of 75 to 125°F. This means that the transmitter gain becomes

$$\frac{(100-0)\%\text{TO}}{(125-75)°\text{F}} = \frac{100\,\%\text{TO}}{50°\text{F}} = 2\frac{\%\text{TO}}{°\text{F}}$$

Thus, the transfer function of the transmitter becomes

$$H = \frac{2.0}{10s+1}$$

and the characteristic equation

$$900s^3 + 420s^2 + 43s + (1+1.6K_C) = 0$$

The new ultimate gain and ultimate period are

$$K_{C_U} = 11.9\frac{\%\text{CO}}{\%\text{TO}} \quad \text{and} \quad T_U = 28.7\,\text{sec}$$

Thus, a change in any gain in the "loop" (in this case we changed the transmitter gain, but any other gain change would have the same effect) will affect K_{C_U}. Furthermore, we can generalize by saying that if any gain in the loop is reduced, K_{C_U} increases. The reciprocal is also true: If any gain in the loop increases, K_{C_U} reduces. The change in gains does not affect the ultimate period.

6-2.2 Effect of Time Constants

Let us now assume that a new faster transmitter (with the same original range of 50 to 150°F) is installed. The time constant of this new transmitter is 5 sec. Thus the transfer function becomes

$$H = \frac{1.0}{5s+1}$$

and the characteristic equation

$$450s^3 + 255s^2 + 38s + (1 + 0.8K_C) = 0$$

The new ultimate gain and ultimate period are

$$K_{C_U} = 25.7 \frac{\%CO}{\%TO} \quad \text{and} \quad T_U = 21.6\,\text{sec}$$

This change in transmitter time constant has affected K_{C_U} and T_U. By reducing the transmitter time constant, K_{C_U} has increased, thus permitting a higher gain before reaching instability, and T_U has been reduced, thus resulting in a faster loop.

   The effect of a change in any time constant cannot be generalized as we did with a change in gain. Again install the original transmitter, and consider now that a change in design results in a faster exchanger; its new transfer function is

$$G_1 = \frac{50}{20s+1}$$

and the characteristic equation

$$450s^3 + 255s^2 + 38s + (1 + 0.8K_C) = 0$$

The new ultimate gain and ultimate period are

$$K_{C_U} = 18.7 \frac{\%CO}{\%TO} \quad \text{and} \quad T_U = 26.8\,\text{sec}$$

The effect of a reduction in the exchanger time constant is completely different from that obtained when the transmitter time constant was changed. In this case, when the time constant was reduced, K_{C_U} also reduced. It is difficult to generalize; however, we can say that by reducing the smaller (nondominant) time constants, K_{C_U} increases, whereas reducing the larger (dominant) time constants, K_{C_U} decreases. Usually, the smaller time constants are those of the instrumentation such as transmitters and valves.

6-2.3 Effect of Dead Time

Back again to the original system, but assume now that the transmitter is relocated to another location farther from the exchanger, as shown in Fig. 6-2.2. This location

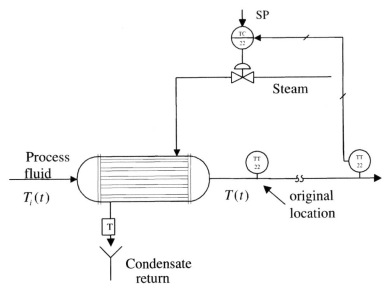

Figure 6-2.2 Heat exchanger showing new transmitter location.

generates a dead time due to transportation. That is, it takes some time to flow from the exchanger to the new transmitter location. Assume that this dead time is only 4 sec. Figure 6-2.3 is a block diagram showing the dead time. The characteristic equation is now

$$900s^3 + 420s^2 + 43s + (1 + 0.8K_C e^{-4s}) = 0$$

The new ultimate gain and ultimate period are

$$K_{C_U} = 9\frac{\%CO}{\%TO} \quad \text{and} \quad T_U = 47.8 \, \text{sec}$$

Note the drastic effect of the dead time on K_{C_U}. A 4-sec dead time has reduced K_{C_U} by 62.2%. T_U was also drastically affected. This proves our comment in Chapter 2 that dead time drastically affects the stability of control loops and therefore the aggressiveness of the controller tunings.

6-2.4   Effect of Integral Action in the Controller

All of the presentation above has been done assuming the controller to be proportional only. A valid question is: How does integration affect K_{C_U} and T_U? Even though Ziegler–Nichols defined the meaning of K_{C_U} for a P controller only, we will still use it because it still is the maximum gain. The transfer function of a PI controller is given by Eq. (3-2.11):

$$G_C = K_C \frac{\tau_I s + 1}{\tau_I s}$$

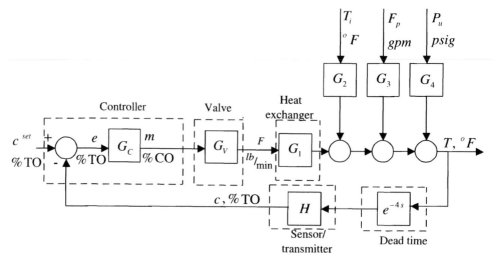

Figure 6-2.3   Block diagram showing dead time.

and the characteristic equation becomes

$$900s^3 + 420s^2 + 43s + \left(1 + 0.8K_C \frac{\tau_I s + 1}{\tau_I s}\right) = 0$$

Using $\tau_I = 30\,\text{sec}$, the ultimate gain and period are

$$K_{CU} = 16.2 \frac{\%CO}{\%TO} \quad \text{and} \quad T_U = 34.4\,\text{sec}$$

Thus *the addition of integration removes the offset, but it reduces K_{CU}.* Integration adds instability to the loop. It also increases T_U, resulting in a slower loop. You may ask yourself: What is the effect of decreasing τ_I? That is, what would happen to K_{CU} if $\tau_I = 20\,\text{sec}$?

6-2.5   Effect of Derivative Action in the Controller

Now that the effect of integration on the loop stability has been studied, what is the effect of the derivative? Let us look at using a PD controller. The transfer function for a PD controller is given by Eq. (3-2.15):

$$G_C = K_C(\tau_D s + 1)$$

and the characteristic equation becomes

$$900s^3 + 420s^2 + 43s + [1 + 0.8K_C(\tau_D s + 1)] = 0$$

Using $\tau_D = 1\,\text{sec}$, the ultimate gain and period are

$$K_{CU} = 36.2 \frac{\%CO}{\%TO} \quad \text{and} \quad T_U = 28.2 \, \text{sec}$$

Thus *the addition of derivative increases the K_{C_U} value, adding stability to the loop!* From a stability point of view, derivative is desirable because it adds stability and therefore makes it possible to tune a controller more aggressively. However, as discussed in Chapter 3, *if noise is present, derivative will amplify it and will be detrimental to the operation.*

In this section we have discussed briefly how to calculate the ultimate gain of a loop. However, we have discussed in more detail how the various gains, time constants, and dead time of a loop affect this ultimate gain. We presented these effects by changing transmitters, process unit (exchanger) design, and so on. What occurs most commonly, however, is that the process unit itself changes, due to its nonlinear characteristics.

6-3 SUMMARY

In this chapter we have presented the development of block diagrams and discussed the important subject of stability of control loops. These subjects are used in all subsequent chapters.

REFERENCE

1. C. A. Smith and A. B. Corripio, *Principles and Practice of Automatic Process Control*, 2nd ed., Wiley, New York, 1997.

CHAPTER 7

FEEDFORWARD CONTROL

In this chapter we present the principles and application of feedforward control, quite often a most profitable control strategy. Feedforward is not a new strategy; the first reports date back to the early 1960s [1,2]. However, the use of computers has contributed significantly to simplify and expand its implementation, which has resulted in increased application of the technique. Feedforward requires a thorough knowledge of the steady-state and dynamics characteristics of the process. Thus good process engineering knowledge is basic to its application.

7-1 FEEDFORWARD CONCEPT

To help us understand the concept of feedforward control, let us briefly recall feedback control; Fig. 7-1.1 depicts the feedback concept. As different disturbances, $D_1(t), D_2(t), \ldots, D_n(t)$, enter the process, the controlled variable $c(t)$ deviates from the set point, and feedback compensates by manipulating another input to the process, the manipulated variable $m(t)$. The advantage of feedback control is its simplicity. Its disadvantage is that to compensate for disturbances, the controlled variable must first deviate from the set point. Feedback acts upon an error between the set point and the controlled variable. It may be thought of as a *reactive* control strategy, since it waits until the process has been upset before it even begins to take corrective action.

By its very nature, feedback control results in a temporary deviation in the controlled variable. Many processes can permit some amount of deviation; however, in many other processes this deviation must be minimized to such an extent that feedback control may not provide the required performance. For these cases, feedforward control may prove helpful.

The idea of feedforward control is to compensate for disturbances before they affect the controlled variable. Specifically, feedforward calls for measuring the

142

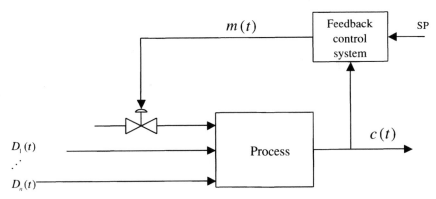

Figure 7-1.1   Feedback control.

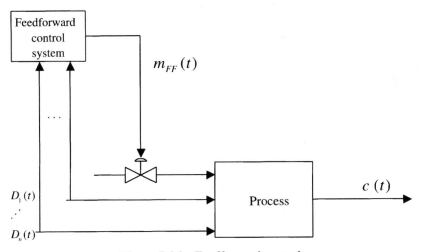

Figure 7-1.2   Feedforward control.

disturbances *before* they enter the process, and based on these measurements, calculating the manipulated variable required to maintain the controlled variable at
set point. If the calculation and action are done correctly, the controlled variable
should remain undisturbed. Thus, feedforward control may be thought of as being
a *proactive* control strategy; Fig. 7-1.2 depicts this concept.

Consider a disturbance $D(t)$, shown in Fig. 7-1.3, entering the process. As soon
as the feedforward controller (FFC) realizes that a change has occurred, it calculates a new value of $m_{FF}(t)$ and sends it to the process (valve). This is done such that
path G_M negates the effect of path G_D.

To attain perfect negation, the feedforward controller must be designed considering the steady-state characteristics of the process. For example, assume that a
change of +1 unit in $D(t)$ affects $c(t)$ by +10 units, and that a change of +1 unit
in $m_{FF}(t)$ affects $c(t)$ by +5 units. Thus, if $D(t)$ changes by +1 unit, affecting $c(t)$ by
+10 units, the feedforward controller must change $m_{FF}(t)$ by −2 units, affecting $c(t)$
by −10 units and therefore negating the effect of $D(t)$.

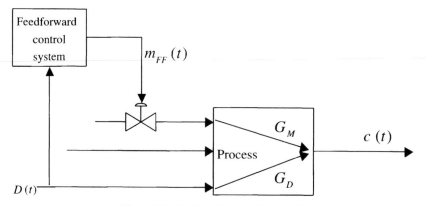

Figure 7-1.3   Feedforward control.

The preceding paragraph explains how the feedforward control strategy compensates considering the steady-state characteristics of the process. However, to avoid any change in the controlled variable, the dynamic characteristics of the process must also be considered. It is desired that the effects of $m_{FF}(t)$ and $D(t)$ affect $c(t)$ at the same time. For example, consider that when $D(t)$ changes, the feed-forward controller changes $m_{FF}(t)$ at almost the same time. If due to process dynamics, the effect of $m_{FF}(t)$ on $c(t)$ is faster than the effect of $D(t)$ on $c(t)$, then $c(t)$ will deviate from its desired value due to $m_{FF}(t)$, not due to $D(t)$. In this case, perfect compensation requires to "slow down" the feedforward controller. That is, the feed-forward controller should not take immediate corrective action; it should wait a certain time before taking action so that the negating effects reach $c(t)$ at the same time. In other processes the effect of $D(t)$ on $c(t)$ may be faster than the effect of $m_{FF}(t)$ on $c(t)$. In these cases perfect compensation requires to "speed up" the feed-forward controller. Thus, *the feedforward controller must be designed to provide the required steady-state and dynamic compensations.*

Figure 7-1.2 shows feedforward compensation for all the disturbances entering the process. However, very often it may be difficult, if not impossible, to measure some disturbances. In addition, some of the possible measurable disturbances may occur infrequently enough that the need for compensation by feedforward may be questionable. Therefore, feedforward control is used to compensate for the "major" measurable disturbances. It is up to the operating personnel to define *major* (disturbances that occur often and/or cause significant deviations in the controlled variable). Feedback control is then used to compensate for those disturbances not compensated by feedforward. Figure 7-1.4 shows the implementation of this feedforward/feedback control. The figure shows that the output of the feedforward controller, $m_{FF}(t)$, and the output of the feedback controller, $m_{FB}(t)$, are added. Section 7-5 shows another way to combine these two outputs.

The above paragraphs have explained the objective, and some design considerations, of feedforward control. The things to keep in mind are that *the feedforward controller contains steady-state and dynamic compensations and that feedback compensation must always be present.*

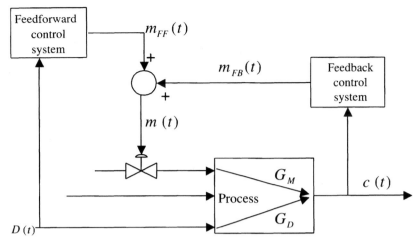

Figure 7-1.4   Feedforward/feedback control.

TABLE 7-2.1   Process Information and Steady-State Values for Mixing Process

Information

Concentration transmitter range = 0.3–0.7 mass fraction. Its dynamic can be described by a time constant of 0.1 min.

The pressure drop across the valve can be considered constant, and the maximum flow provided by the valve is 3800 gpm. The valve dynamics can be described by a time constant of 0.1 min.

The densities of all streams are also considered similar and constant.

Steady-State Values

Stream	Flow (gpm)	Mass Fraction	Stream	Flow (gpm)	Mass Fraction
1	1900	0.000	5	500	0.800
2	1000	0.990	6	3900	0.472
3	2400	0.167	7	500	0.900
4	3400	0.049			

7-2   BLOCK DIAGRAM DESIGN OF LINEAR FEEDFORWARD CONTROLLERS

In the present section and the following three sections we show the design of feedforward controllers. The mixing system shown in Fig. 7-2.1 is used to illustrate this design; Table 7-2.1 gives the steady-state conditions and other process information. In this process three different streams, each composed of component A and water, are mixed and diluted with water to a final desired composition of component A, $x_6(t)$. Due to process considerations the mixing must be done in three constant-volume tanks as shown in the figure. All of the input streams represent possible disturbances to the process; that is, the flows and compositions of streams 5, 2, and 7

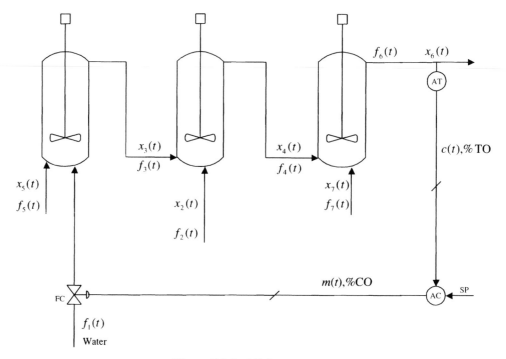

Figure 7-2.1 Mixing process.

may vary. However, the major disturbances usually come from stream 2. Commonly, the stream flow $f_2(t)$ may double, while the mass fraction, $x_2(t)$, may drop as much as 20% of its steady-state value. Figure 7-2.2 shows the control provided by feedback control when $f_2(t)$ changes from 1000 gpm to 2000 gpm. The composition changes from its steady-state value of 0.472 mass fraction (mf) to about 0.525 mf, a 11.22% change from set point. The maximum deviation permitted for this process is, however, only 1.5% from set point. That is, the maximum value of composition permitted is 0.479 mf, and the minimum value permitted is 0.465 mf. Thus it does not appear that simple feedback can provide the required performance; feedforward control may be justified.

Assuming for the moment that $f_2(t)$ is the major disturbance, the application of feedforward to this process calls for measuring this flow, and on a change take corrective action. The design of this feedforward controller now follows.

To begin, let us draw the block diagram for the process; this is shown in Fig. 7-2.3. The diagram shows $f_2(t)$ as the disturbance of concern. The significance of each transfer function is as follows:

G_C = transfer function of the composition controller

G_V = transfer function of the valve; describes how the controller's output affects the water flow

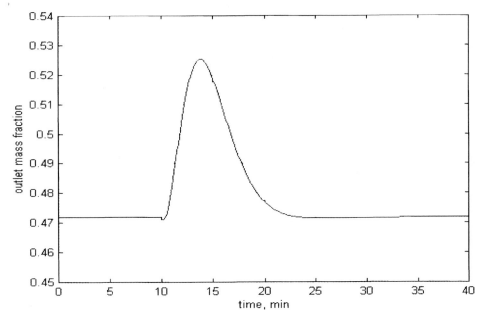

Figure 7-2.2   Feedback control when $f_2(t)$ changes from 1000 gpm to 2000 gpm.

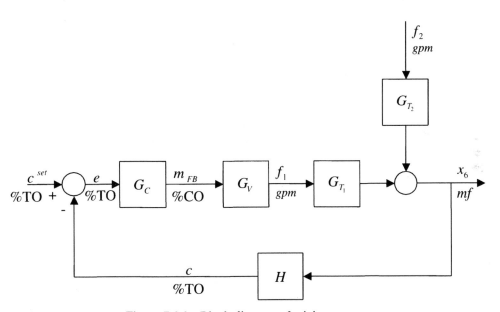

Figure 7-2.3   Block diagram of mixing process.

G_{T_1} = transfer function of the mixing process; describes how the water flow affects the outlet composition

G_{T_2} = transfer function of the mixing process; describes how $f_2(t)$ affects the outlet composition

H = transfer function of the composition sensor and transmitter

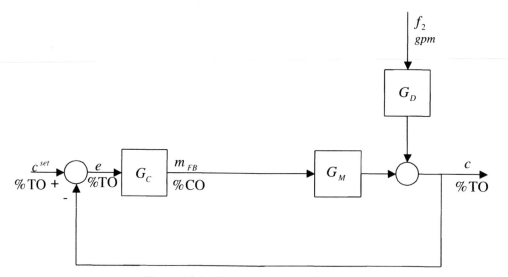

Figure 7-2.4   Condensed block diagram.

A more condensed block diagram, shown in Fig. 7-2.4, can now be drawn. The significance of each transfer function is as follows:

G_M = transfer function that describes how the manipulated variable, $m(t)$, affects the controlled variable, $c(t)$ (in this case, $G_M = G_V G_{T_1} H$)

G_D = transfer function that describes how the disturbance, $f_2(t)$, affects the controlled variable (in this case, $G_D = G_{T_2} H$)

To review, the objective of feedforward control is to measure the inputs, and if a change is detected, adjust the manipulated variable to maintain the controlled variable at set point. This control operation is shown in Fig. 7-2.5 and the block diagram in Fig. 7-2.6. The significance of each new transfer function is as follows:

H_D = transfer function that describes the sensor and transmitter that measures the disturbance

FFC = transfer function of feedforward controller

Note that in Figs. 7-2.5 and 7-2.6 the feedback controller has been "disconnected." This controller will be "connected" again later.

Figure 7-2.6 shows that the way a change in disturbance, Δf_2, affects the controlled variable is given by

$$\Delta c = G_D \Delta f_2 + H_D(\text{FFC})G_M \Delta f_2$$

The objective is to design FFC such that a change in $f_2(t)$ does not affect $c(t)$, that is, such that $\Delta c = 0$. Thus

$$0 = G_D \Delta f_2 + H_D(\text{FFC})G_M \Delta f_2$$

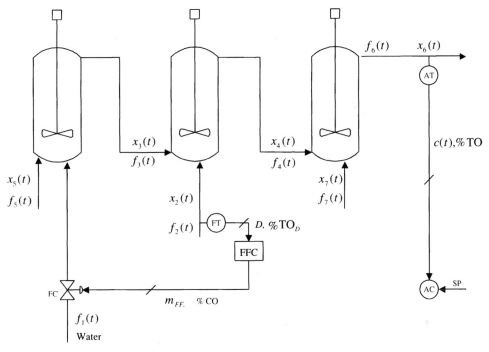

Figure 7-2.5   Feedforward control system.

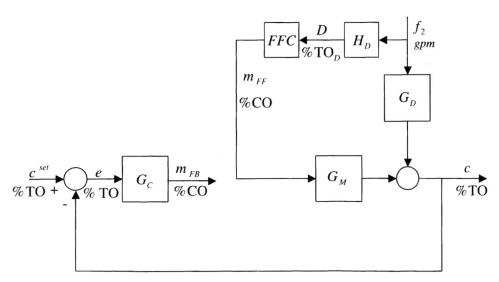

Figure 7-2.6   Block diagram of feedforward control system.

Dividing both sides by Δf_2 and solving for FFC yields

$$\text{FFC} = -\frac{G_D}{H_D G_M} \qquad (7\text{-}2.1)$$

Equation (7-2.1) is the *design formula for the feedforward controller.* We understand that at this moment, this design formula does not say much; furthermore, you wonder what is it all about. Don't despair, let us give it a try.

As learned in earlier chapters, first-order-plus-dead-time transfer functions are commonly used as an approximation to describe processes; Chapter 2 showed how to evaluate this transfer function from step inputs. Using this type of approximation for this process,

$$G_D = \frac{K_D e^{-t_{0_D} s}}{\tau_D s + 1} \qquad \frac{\%\text{TO}}{\text{gpm}} \qquad (7\text{-}2.2)$$

$$G_M = \frac{K_M e^{-t_{0_M} s}}{\tau_M s + 1} \qquad \frac{\%\text{TO}}{\%\text{CO}} \qquad (7\text{-}2.3)$$

and assuming that the flow transmitter is very fast, H_D is only a gain:

$$H_D = K_{T_D} \qquad \frac{\%\text{TO}_D}{\text{gpm}} \qquad (7\text{-}2.4)$$

Substituting Eqs. (7-2.2), (7-2.3), and (7-2.4) into (7-2.1) yields

$$\text{FFC} = -\frac{K_D}{K_{T_D} K_M} \frac{\tau_M s + 1}{\tau_D s + 1} e^{-(t_{0_D} - t_{0_M})s} \qquad (7\text{-}2.5)$$

We next explain in detail each term of this feedforward controller.

The first element of the feedforward controller, $-K_D/K_{T_D}K_M$, contains only gain terms. This term is the part of the feedforward controller that compensates for the steady-state differences between the G_D and G_M paths. The units of this term help in understanding its significance:

$$\frac{K_D}{K_{T_D} K_M} [=] \frac{\%\text{TO}/\text{gpm}}{(\%\text{TO}_D/\text{gpm})/(\%\text{TO}/\%\text{CO})} = \frac{\%\text{CO}}{\%\text{TO}_D}$$

Thus the units show that the term indicates how much the feedforward controller output, $m_{\text{FF}}(t)$ in %CO, changes per unit change in transmitter's output, D in %TO_D.

Note the minus sign in front of the gain term. This sign helps to decide the "action" of the controller. In the process at hand, K_D is positive, because as $f_2(t)$ increases, the outlet concentration $x_6(t)$ also increases because stream 2 is more concentrated than the outlet stream. K_M is negative, because as the signal to the valve increases, the valve opens, more water flow enters, and the outlet concentration decreases. Finally, K_{T_D} is positive, because as $f_2(t)$ increases, the signal from the transmitter also increases. Thus the sign of the gain term is negative:

$$\frac{K_D}{K_{T_D} K_M} \rightarrow \frac{+}{+\,-} = -$$

A negative sign means that as $\%TO_D$ increases, indicating an increase in $f_2(t)$, the feedforward controller output $m_{FF}(t)$ should decrease, closing the valve. This action does not make sense. As $f_2(t)$ increases, tending to increase the concentration of the output stream, the water flow should also increase, to dilute the outlet concentration, thus negating the effect of $f_2(t)$. Therefore, the sign of the gain term should be positive. Notice that when the negative sign in front of the gain term is multiplied by the sign of this term, it results in the correct feedforward action. Thus the negative sign is an important part of the controller.

The second term of the feedforward controller includes only the time constants of the G_D and G_M paths. This term, referred to as *lead/lag*, compensates for the differences in time constants between the two paths. In Section 7-3 we discuss this term in detail.

The last term of the feedforward controller contains only the dead-time terms of the G_D and G_M paths. This term compensates for the differences in dead time between the two paths and is referred to as a *dead-time compensator*. Sometimes the term $t_{o_D} - t_{o_M}$ may be negative, yielding a positive exponent. As we learned in Chapter 2, the Laplace representation of dead time includes a negative sign in the exponent. When the sign is positive, it is definitely not a dead time and cannot be implemented. A negative sign in the exponential is interpreted as "delaying" an input; a positive sign may indicate a "forecasting." That is, the controller requires taking action before the disturbance happens. This is not possible. When this occurs, quite often there is a physical explanation, as the present example shows.

Thus it can be said that the first term of the feedforward controller is a steady-state compensator, while the last two terms are dynamic compensators. All these terms are easily implemented using computer control software; Fig. 7-2.7 shows the implementation of Eq. (7-2.5). Years ago, when analog instrumentation was solely used, the dead-time compensator was either extremely difficult or impossible to implement. At that time, the state of the art was to implement only the steady-state and lead/lag compensators. Figure 7-2.6 shows a component for each calculation needed for the feedforward controller, that is, one component for the dead time, one for the lead/lag, and one for the gain. Very often, however, lead/lags have adjustable gains, and in this case we can combine the lead/lag and gain into only one component.

Well, enough for this bit of theory, and let us see what results out of all of this. Returning to the mixing system, under open-loop conditions, a step change of 5% in the signal to the valve provides a process response form where the following first-order-plus-dead-time approximation is obtained:

$$G_M = \frac{-1.095e^{-0.9s}}{3.50s+1} \qquad \frac{\%TO}{\%CO} \qquad (7\text{-}2.6)$$

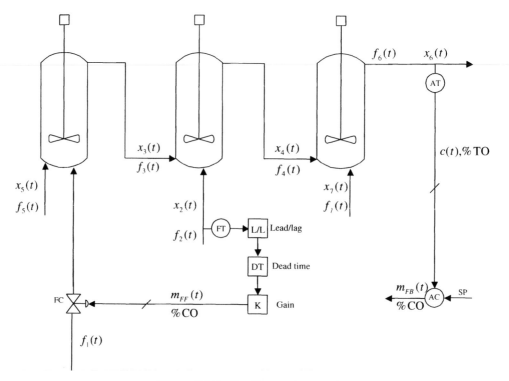

Figure 7-2.7 Feedforward control.

Also under open-loop conditions, $f_2(t)$ was allowed to change by 10 gpm in a step fashion, and from the process response the following approximation is obtained:

$$G_D = \frac{0.032e^{-0.75s}}{2.75s+1} \qquad \frac{\%TO}{gpm} \tag{7-2.7}$$

Finally, assuming that the flow transmitter in stream 2 is calibrated from 0 to 3000 gpm, its transfer function is given by

$$H_D = \frac{100\%TO_D}{3000\,gpm} = 0.033\,\frac{\%TO_D}{gpm} \tag{7-2.8}$$

Substituting the previous three transfer functions into Eq. (7-2.5) yields

$$FFC = 0.891\left(\frac{3.50s+1}{2.75s+1}\right)e^{-(0.75-0.9)s}$$

The dead time indicated, 0.75 to 0.9, is negative and therefore the dead-time compensator cannot be implemented. Thus the implementable, or realizable, feedforward controller is

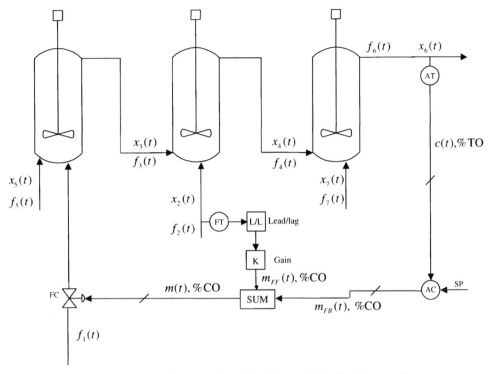

Figure 7-2.8   Implementation of feedforward/feedback controller.

$$\text{FFC} = 0.891\left(\frac{3.50s+1}{2.75s+1}\right) \tag{7-2.9}$$

Figure 7-2.8 shows the implementation of this controller. The figure shows that the feedback compensation has also been implemented. This implementation has been accomplished by adding the output of both feedforward and feedback controllers using a summer. Section 7-4 discusses how to implement this addition. Figure 7-2.9 shows the block diagram for this combined control scheme.

Figure 7-2.10 shows the response of the composition when $f_2(t)$ doubles from 1000 gpm to 2000 gpm. The figure compares the control provided by feedback control (FBC), steady-state feedforward control (FFCSS), and dynamic feedforward control (FFCDYN). In steady-state feedforward control, no dynamic compensation is implemented; that is, in this case the feedforward controller is FFC = 0.891. Dynamic feedforward control includes the complete controller, Eq. (7-2.9). Under steady-state feedforward the mass fraction increased up to 0.477 mf, a 1.05% change from the set point. Under dynamic feedforward the mass fraction increased up to 0.473 mf, or 0.21%. The improvement provided by feedforward control is quite impressive. Figure 7-2.10 also shows that the process response tends to decrease first and then increase; we discuss this response later.

The previous paragraphs and figures have shown the development of a linear feedforward controller and the responses obtained. At this stage, since we have not yet offered an explanation of the lead/lag unit, the reader may be wondering about it. Let us explain this term before further discussing feedforward control.

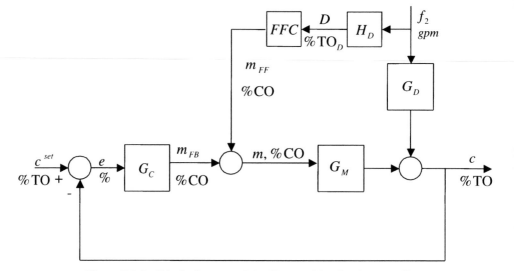

Figure 7-2.9   Block diagram of feedforward/feedback controller.

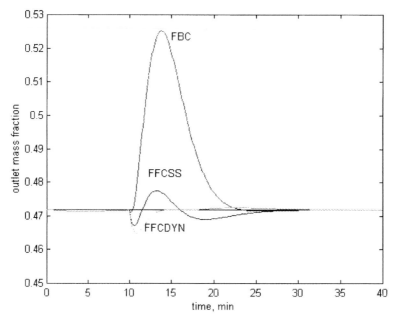

Figure 7-2.10   Feedforward and feedback responses when $f_2(t)$ changes from 1000 gpm to 2000 gpm.

7-3   LEAD/LAG TERM

As indicated in Eqs. (7-2.5) and (7-2.9), the lead/lag term is composed of a ratio of two $\tau s + 1$ terms; or more specifically, its transfer function is given by

$$\frac{O(s)}{I(s)} = \frac{\tau_{ld}s + 1}{\tau_{lg}s + 1} \tag{7-3.1}$$

where $O(s)$ is the Laplace transform of output variable, $I(s)$ the Laplace transform of input variable, τ_{ld} the lead time constant, and τ_{lg} the lag time constant.

To explain the workings of the lead/lag term let us suppose that the input changes, in a step fashion, with A units of amplitude. The equation that describes how the output responds to this input is

$$O(t) = A\left(1 + \frac{\tau_{ld} - \tau_{lg}}{\tau_{lg}}e^{-t/\tau_{lg}}\right) \tag{7-3.2}$$

Figure 7-3.1 shows the response for different values of the ratio τ_{ld}/τ_{lg} while keeping $\tau_{lg} = 1$; the input is a step change of 5 units of magnitude. The figure shows that as the ratio increases, the initial response also increases; as time increases, the response approaches asymptotically its final value of 5 units. For values of $\tau_{ld}/\tau_{lg} > 1$ the initial response (equal to the input change times the ratio) at $t = 0$ is greater than its final value, while for values of $\tau_{ld}/\tau_{lg} < 1$ the initial response (also equal to the input change times the ratio) is less than its final value. Therefore, the initial response depends on the ratio of the lead time constant to the lag time constant, τ_{ld}/τ_{lg}. The time approach to the final value depends only on the lag time constant,

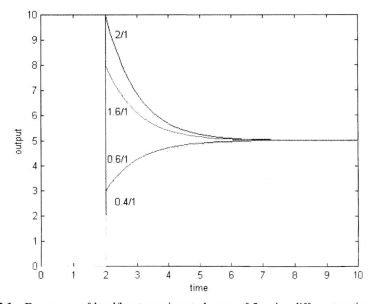

Figure 7-3.1   Response of lead/lag to an input change of 5 units, different ratios of τ_{ld}/τ_{lg}.

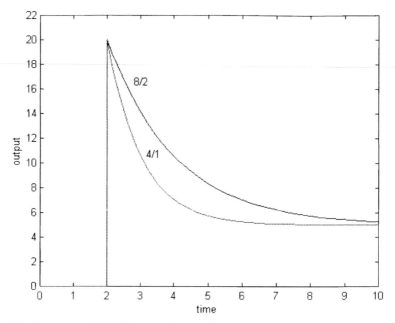

Figure 7-3.2   Response of lead/lag to an input change of 5 units, different ratios of τ_{ld}/τ_{lg}.

τ_{lg}. Thus, in tuning a lead/lag, both τ_{ld} and τ_{lg} must be provided. The reader should use Eq. (7-3.2) to convince himself or herself of what was just explained.

Figure 7-3.2 is shown to further help in understanding lead/lags. The figure shows two response curves with identical values of the ratio τ_{ld}/τ_{lg} but different individual values of τ_{ld} and τ_{lg}. The figure shows that the magnitude of the initial output response is the same, because the ratio is the same, but the response with the larger τ_{lg} takes longer to reach the final value.

Equation (7-2.5) indicates the use of a lead/lag term in the feedforward controller. The equation indicates that τ_{ld} should be set equal to τ_M and that τ_{lg} should be set equal to τ_D.

7-4   EXTENSION OF LINEAR FEEDFORWARD CONTROLLER DESIGN

With an understanding of the lead/lag term, we can now return to the example of Section 7-2: specifically, to a discussion of the dynamic compensation of the feedforward controller. Comparing the transfer functions given by Eqs. (7-2.6) and (7-2.7), it is easy to realize that the controlled variable $c(t)$ responds slower to a change in the manipulated variable $m(t)$ than to a change in the disturbance $f_2(t)$. Recall that a design consideration for a feedforward controller is to compensate for the dynamic differences between the manipulated and the disturbance paths, the G_D and G_M paths. The feedforward controller for this process should be designed to speed up the response of the controlled variable on a change in the manipulated variable. That is, the feedforward controller should speed up the G_M path; the resulting controller, Eq. (7-2.9), does exactly this. First, note that the resulting lead/lag term has a τ_{ld}/τ_{lg} ratio greater than 1, $\tau_{ld}/\tau_{lg} = 3.50/2.75 = 1.27$. This means that at the moment the signal from the flow transmitter changes by 1%, indicating a certain change in $f_2(t)$, the lead/lag output changes by 1.27%, resulting in an initial output change from the feedforward controller of $(0.891)(1.27) = 1.13\%$. Eventually, the lead/lag output approaches 1%, and the feedforward controller output approaches 0.891%. This type of action results in an initial increase in $f_1(t)$ greater than the one really needed for the specific increase in $f_2(t)$. This initial greater increase provides a "kick" to the G_M path to move faster, resulting in tighter control than in the control provided by steady-state feedforward control, as shown in Fig. 7-2.10. Second, note that the feedforward controller equation does not contain a dead-time term. There is no need to delay the feedforward action. On the contrary, the present process requires us to speed up the feedforward action. Thus the absence of a dead-time term makes sense.

It is important to realize that this feedforward controller, Eq. (7-2.9), only compensates for changes in $f_2(t)$. Any other disturbance will not be compensated by the feedforward controller, and in the absence of a feedback controller it would result in a deviation of the controlled variable. The implementation of feedforward control requires the presence of feedback control. Feedforward control compensates for the major measurable disturbances, while feedback control takes care of all other disturbances. In addition, any inexactness in the feedforward controller is also compensated by the feedback controller. Thus *feedforward control must be implemented with feedback compensation.* Feedback from the controlled variable must be present.

Figure 7-2.7 shows a summer where the signals from the feedforward controller $m_{FF}(t)$ and from the feedback controller $m_{FB}(t)$ are combined. The summer solves the equation

$$m(t) = \text{feedback signal} + \text{feedforward signal} + \text{bias}$$

To be more specific,

$$m(t) = K_X X + K_Y Y + B$$

Let the feedback signal be the X input, the feedforward signal the Y input. Therefore,

$$m(t) = K_X m_{FB}(t) + K_Y m_{FF}(t) + B$$

As discussed previously, the sign of the steady-state part of the feedforward controller is positive for this process. Thus the value of K_Y is set to +1; if the sign had been negative, K_Y would have been set to –1. The value of K_X is also set to +1. Note that by setting K_Y to 0 or to 1 provides an easy way to turn the feedforward controller on or off.

Let us suppose that the process is at steady state under feedback control only ($K_Y = B = 0$, $K_X = 1$) and it is now desired to turn the feedforward controller on. Furthermore, since the process is at steady state, it is desired to turn the feedforward controller on without upsetting the signal to the valve. That is, a "bumpless transfer" from simple feedback control to feedforward/feedback control is desired. To accomplish this transfer, the summer is first set to manual, which freezes its output, K_Y is set to +1, the output of the feedforward controller is read from the output of the gain block, the bias term B is set equal to the negative of the value read, and finally, the summer is set back to automatic. This procedure results in the bias term canceling the feedforward controller output. To be a bit more specific, suppose that the process is running under feedback control only, with a signal to the valve equal to $m_{FB}(t)$. It is then desired to "turn on" the feedforward controller, and at this time the process is at steady state with $f_2(t) = 1500$ gpm. Under this condition the output of the flow transmitter is at 50%, and $m_{FF}(t) = 0.891 \times 50 = 44.55\%$. Then the procedure just explained is followed, yielding

$$m(t) = (1)m_{FB}(t) + (1)(44.55) - 44.55 = m_{FB}(t)$$

Now suppose that $f_2(t)$ changes from 1500 gpm to 1800 gpm, making the output from the flow transmitter equal to 60%. After the transients through the lead/lag have died out, the output from the feedforward controller becomes equal to 53.46%. Thus, the feedforward controller asks for 8.91% more signal to the valve to compensate for the disturbance. At this moment, the summer output signal is

$$m(t) = (1)m_{FB}(t) + (1)(53.46) - 44.55 = m_{FB}(t) + 8.91\%$$

which changes the signal to the valve by the amount required.

The procedure just described to implement the summer is easy; however, it requires manual intervention by the operating personnel. Most control systems can easily be configured to perform the procedure automatically. For instance, consider the use of an on–off switch and two bias terms, B_{FB} and B_{FF}. The switch is used to indicate only feedback control (switch is OFF) or feedforward (switch is ON). B_{FB} is used when only feedback control is used ($K_Y = 0$), and B_{FF} is used when feedforward control is used ($K_Y = 1$).

$$m(t) = K_X m_{FB}(t) + K_Y m_{FF}(t) + (B_{FB} \text{ if switch is OFF or } B_{FF} \text{ if switch is ON})$$

$$(7\text{-}4.1)$$

Originally, $B_{FB} = B_{FF} = 0$. While only feedback is used, the following is being calculated:

$$B_{FF} = -m_{FF}(t) + B_{FB} \tag{7-4.2}$$

As soon as the switch goes ON, this calculation stops and B_{FF} remains constant at the last value calculated. While feedforward is being used, the following is being calculated:

$$B_{FB} = m_{FF}(t) + B_{FF} \tag{7-4.3}$$

As soon as the switch goes OFF, this calculation stops and B_{FB} remains constant at the last value calculated. This procedure guarantees automatic bumpless transfer. The reader is encouraged to test this algorithm.

In the previous paragraphs we have explained just one way to implement the summer where the feedback and feedforward signals are combined. The importance of bumpless transfer was stressed. The way the summer is implemented depends on the algorithms provided by the control system used. For example, there are control systems that provide a lead/lag and a summer in only one algorithm, called a *lead/lag summer*. In this case the feedback signal can be brought directly into the lead/lag, and summation is done in the same unit; the summer unit is not needed. There are other control systems that provide what they call a PID feedforward. In this case the feedforward signal is brought into the feedback controller and is added to the feedback signal calculated by the controller. How the bumpless transfer is accomplished depends on the control system.

In the example presented so far, feedforward control has been implemented to compensate for $f_2(t)$ only. But what if it is necessary to compensate for another disturbance, such as $x_2(t)$? The technique to design this new feedforward controller is the same as before; Fig. 7-4.1 shows a block diagram including the new disturbance with the new feedforward controller FFC$_2$.

Following the previous development, the new controller equation is

$$\text{FFC}_2 = -\frac{G_{D_2}}{H_{D_2}G_M} \tag{7-4.4}$$

Step testing the mass fraction of stream 2 yields the following transfer function:

$$G_{D_2} = \frac{63.87e^{-0.85s}}{2.5s+1}; \quad \frac{\%\text{TO}}{\text{mf}} \tag{7-4.5}$$

Assuming that the concentration transmitter in stream 2 has a negligible lag and that it has been calibrated from 0.5 to 1.0 mf, its transfer function is given by

$$H_{D_2} = \frac{100\%\text{TO}_{D_2}}{0.5\,\text{mf}} = 200\frac{\%\text{TO}_{D_2}}{\text{mf}} \tag{7-4.6}$$

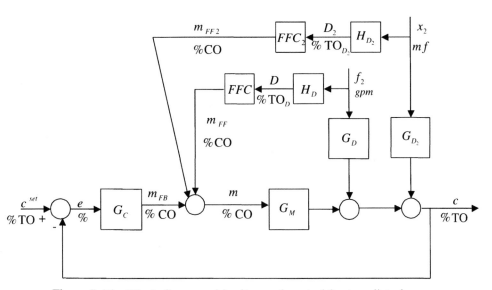

Figure 7-4.1 Block diagram of feedforward control for two disturbances.

Finally, substituting Eqs. (7-2.6), (7-4.5), and (7-4.6) into Eq. (7-4.4) yields

$$FFC_2 = 0.293\left(\frac{3.00s+1}{2.50s+1}\right)e^{-(0.85-0.9)s} \qquad (7\text{-}4.7)$$

Because the dead time is again negative,

$$FFC_2 = 0.293\left(\frac{3.00s+1}{2.50s+1}\right) \qquad (7\text{-}4.8)$$

Figure 7-4.2 shows the implementation of this new feedforward controller added to the previous one and to the feedback controller. Figure 7-4.3 shows the response of $x_6(t)$ to a change of $-0.2\,mf$ in $x_2(t)$ under feedback control, steady-state feedforward, and dynamic feedforward control. The improvement provided by feedforward control is certainly significant. Most of the improvement in this case is provided by the steady-state term; the addition of lead/lag provides an arguably improvement. It is a judgment call in this case whether or not to implement lead/lag. Note that the ratio of the lead-time constant to the lag-time constant is 1.20, which is close to 1.0. Based on our discussion of the lead/lag term, the closer the ratio is to 1.0, the less the need for lead/lag compensation. *A rule of thumb* that could be used to decide whether or not to use lead/lag is: If τ_{ld}/τ_{lg} is between 0.75 and 1.25,

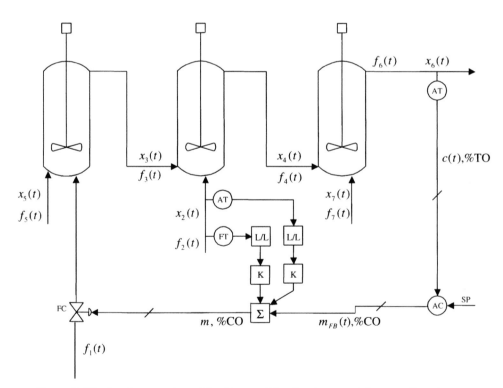

Figure 7-4.2 Implementation of feedforward/feedback control for two disturbances.

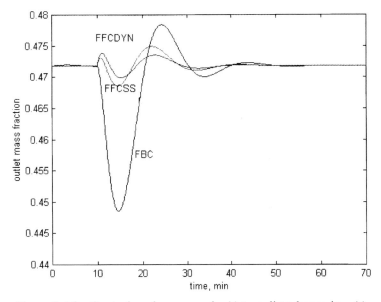

Figure 7-4.3   Control performance of $x_6(t)$ to a disturbance in $x_2(t)$.

do not use lead/lag. The reason for this rule is because the added complexity hardly affects the results. Outside these limits the use of lead/lag may significantly improve the control performance.

When more than one disturbance is compensated by feedforward, the algorithm used to sum the feedforward and feedback signals must be expanded. Specifically, Eq. (7-4.2) becomes

$$B_{FF} = -\sum \left[m_{FF_i}(t) + B_{FB} \right] \qquad (7\text{-}4.9)$$

and Eq. (7-4.3) becomes

$$B_{FB} = \sum \left[m_{FF_i}(t) + B_{FF} \right] \qquad (7\text{-}4.10)$$

7-5   DESIGN OF NONLINEAR FEEDFORWARD CONTROLLERS FROM BASIC PROCESS PRINCIPLES

There are two important considerations of the feedforward controllers developed thus far, Eqs. (7-2.9) and (7-4.8). First, both controllers are linear; they were developed from linear models of the process which are valid only for small deviations around the operating point where the step tests were performed. These controllers are then used with the same constant parameters without consideration of the operating conditions. As learned in Chapter 2, processes most often have nonlinear characteristics; consequently, as operating conditions change, the control performance provided by linear controllers may degrade.

The second consideration is that step changes in the manipulated variable and in the disturbance(s) are required. Often, step changes in the disturbances are not obtained easily. For example, how would you insert a step change in $x_2(t)$ to obtain Eq. (7-4.5)? Certainly, this in not easy and may not be possible.

As discussed in Section 7-1, feedforward controllers are composed of steady-state and dynamic compensators. Very often, the steady-state compensator, which we have called $-K_D/K_{T_D}K_M$, can be obtained by other means, yielding a nonlinear compensator and not requiring step changes in variables. The nonlinear compensator provides an improved control performance over a wide range of operating conditions.

One method to obtain a nonlinear steady-state compensator consists of using first principles, usually mass or energy balances. Using first principles, it is desired to develop an equation that provides the manipulated variable as a function of the disturbances and the set point of the controlled variable. That is,

$$m(t) = f[d_1(t), d_2(t), \ldots, d_n(t), \text{setpoint}]$$

For the process at hand,

$$f_1(t) = f[f_5(t), x_5(t), f_2(t), x_2(t), f_7(t), x_7(t), x_6^{\text{set}}(t)]$$

where $x_6^{\text{set}}(t)$ is the set point of $x_6(t)$.

In Section 7-4 we decided that for this process the major disturbances are $f_2(t)$ and $x_2(t)$ and that the other inlet flows and compositions are minor disturbances. Thus, we need to develop an equation, the steady-state feedforward controller, that relates the manipulated variable $f_1(t)$ in terms of the disturbances $f_2(t)$ and $x_2(t)$. In this equation we consider all other inlet flows and compositions at their steady-state values. That is,

$$f_1(t) = f[\bar{f}_5, \bar{x}_5, f_2(t), x_2(t), \bar{f}_7, \bar{x}_7, x_6^{\text{set}}(t)]$$

where the overbar indicates the steady-state values of the variables.

Because we are dealing with compositions and flows, mass balances are the appropriate first principles to use. Since there are two components, A and water, we can write two independent mass balances. We start with a total mass balance around the three tanks:

$$\rho \bar{f}_5 + \rho f_1(t) + \rho f_2(t) + \rho \bar{f}_7 - \rho f_6(t) = 0 \quad \text{one equation, two unknowns} [f_1(t), f_6(t)]$$
$$(7\text{-}5.1)$$

Note that $f_2(t)$ is not considered an unknown because it will be measured and thus its value will be known. A mass balance on component A provides the other equation:

$$\rho \bar{f}_5 \bar{x}_5 + \rho f_2(t)x_2(t) + \rho \bar{f}_7 \bar{x}_7 - \rho f_6(t)x_6^{\text{set}}(t) = 0 \quad \text{two equations, two unknowns}$$
$$(7\text{-}5.2)$$

Because $x_2(t)$ will also be measured, it is not considered an unknown. Solving for $f_6(t)$ from Eq. (7-5.1), substituting into Eq. (7-5.2), and rearranging yields

$$f_1(t) = \frac{1}{x_6^{set}(t)}(\bar{f}_5\bar{x}_5 + \bar{f}_7\bar{x}_7) - \bar{f}_5 - \bar{f}_7 + \frac{1}{x_6^{set}(6)}[x_2(t) - x_6^{set}(t)]f_2(t) \qquad (7\text{-}5.3)$$

Substituting the steady-state values into Eq. (7-5.3) yields

$$f_1(1) = \frac{1}{x_6^{set}(t)}[850 + f_2(t)x_2(t)] - f_2(t) - 1000 \qquad (7\text{-}5.4)$$

Equation (7-5.4) is the desired steady-state feedforward controller.

The implementation of Eq. (7-5.4) depends on how the feedback correction, the output of the feedback controller, is implemented. This implementation depends on the physical significance given to the feedback signal; there are several ways to do so.

One way is to decide that the significance of the feedback signal is $\Delta f_1(t)$ and use a summer similar to that in Fig. 7-4.2. In this case we first substitute $x_6^{set}(t) = 0.472$ into Eq. (7-5.4) and obtain

$$f_1(t) = 800.85 + f_2(t)\left[\frac{x_2(t)}{0.472} - 1\right] \qquad (7\text{-}5.5)$$

This equation is written in engineering units. Depending on the control system being used, the equation may have to be scaled before implemented. Assuming that this is done, if needed, Fig. 7-5.1 shows the implementation of this controller; a multiplier is needed only with no dynamic compensation. Please note that because Eq. (7-5.5) provides $f_1(t)$, a flow loop has been added to stream 1. If it is decided not to use the flow loop, a conversion between $f_1(t)$ and the valve position should be inserted in Eq. (7-5.5).

Another way to implement the feedback compensation is by deciding that the significance of the feedback signal is $1/x_6^{set}$. This signal is then input into Eq. (7-5.4) to calculate $f_1(t)$. Thus, *in this case the feedback signal is used directly in the feedforward calculation* and not to bias it; Fig. 7-5.2 shows the implementation of this controller. The figure shows only one block referred to as CALC. The actual number of computing blocks, or software, required to implement Eq. (7-5.4) depends on the control system used.

Figure 7-5.3 shows the response of the process under feedback controller and the two nonlinear steady-state feedforward controllers to disturbances of a 500-gpm decrease in $f_2(t)$ and a 0.2-mf decrease in $x_2(t)$. The response FFCNL1 is obtained when Eq. (7-5.5) is used (Fig. 7-5.1). The response FFCNL2 is obtained when Eq. (7-5.4) is used (Fig. 7-5.2). The improvement in control performance obtained with the nonlinear controllers is obvious. The improved performance obtained with the second nonlinear controller is quite impressive. This controller describes more accurately the nonlinear characteristics of the process and can provide better control.

Instead of calling the output of the feedback controller $1/x_6^{set}$, we could have alternatively called it x_6^{set}. The control performance would be the same, but what about the action of the feedback controller in both cases? Think about it.

Previous paragraphs have shown two different ways to implement the nonlinear

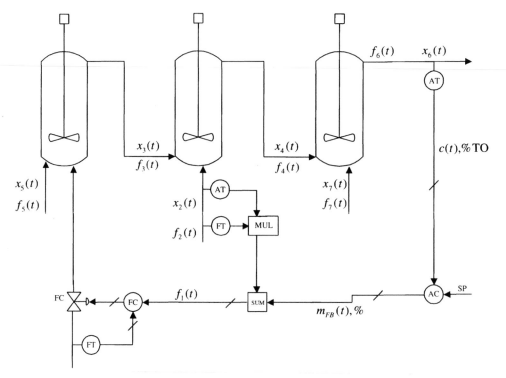

Figure 7-5.1   Nonlinear feedforward control.

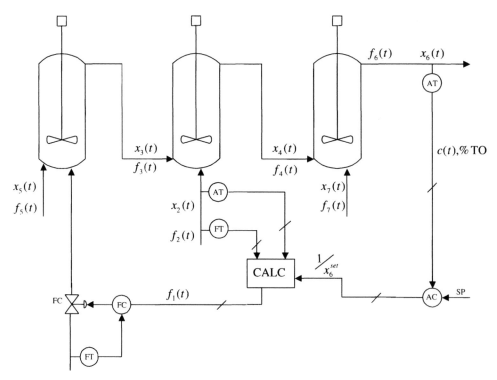

Figure 7-5.2   Nonlinear feedforward control.

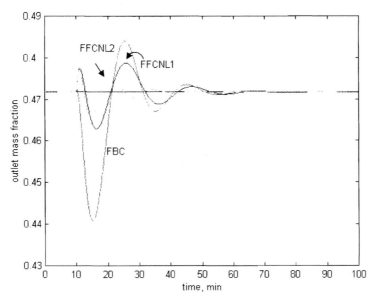

Figure 7-5.3   Control performance of $x_6(t)$ to a decrease of 500 gpm in $f_2(t)$ and a decrease of 0.2 in $x_2(t)$.

steady-state feedforward controller, depending on the significance given to the feedback signal. The designer has complete freedom to make this decision. In the first case the feedback controller biased the feedforward calculation. This is a simple and valid choice and the one most commonly used. The second choice is also a simple choice. Please note that the actual desired value of x_6^{set} is the set point to the feedback controller. The feedback controller changes the term $1/x_6^{set}$, or x_6^{set}, in the feedforward equation to keep its own set point.

Sometimes the development of a nonlinear steady-state compensator from first principles may be just difficult to obtain. Fortunately, process engineering tools provide yet another way to develop this compensator. Processes are usually designed either by steady-state flowsheet simulators or any other steady-state simulation. These simulators, together with regression analysis tools, provide another avenue to design the steady-state compensator. The simulation can be run at different conditions [i.e., different $f_2(t)$, $x_2(t)$, and x_6^{set}] and the required manipulated variable $f_1(t)$ can be calculated to keep the controlled variable at set point. This information can then be fed to a multiple regression program to develop an equation relating the manipulated variables to the disturbances and set point.

7-6   CLOSING COMMENTS ON FEEDFORWARD CONTROLLER DESIGN

There are some comments about the process and example presented in this section, and about feedforward control in general, that should be discussed before proceeding with more examples.

The first comment refers to the process itself. Figure 7-2.10 shows the response of the control system when $f_2(t)$ changes from 1000 gpm to 2000 gpm. The composition of this stream is quite high (0.99), and thus this change in $f_2(t)$ tends to increase $x_6(t)$. However, the response shown in Fig. 7-2.10 shows that initially the composition $x_6(t)$ tends to decrease and then increase. This behavior is an *inverse response*, and of course there is an explanation for this behavior. Because the tanks are at constant volume, an increase in $f_2(t)$ results in an immediate increase in $f_4(t)$. The composition of stream 4, which enters the third tank, is less than the composition of stream 6, which exits the third tank. Thus this increase in $f_4(t)$ tends initially to dilute (decrease) the composition $x_6(t)$. Eventually, the increase in $f_2(t)$ results in an increase in the composition entering the third tank and a corresponding increase in $x_6(t)$. Figure 7-2.10 shows that the response under feedforward control exhibits a more pronounced inverse response. What occurs is that when $f_2(t)$ increases, $f_1(t)$ is also increased by the feedforward controller. Thus the total flow to the third tank increases even more, and the dilution effect in that tank is more pronounced. Could the reader explain why the inverse response is more pronounced under dynamic feedforward than under steady-state feedforward?

The second comment refers to the lead/lag term. The lead/lag is a simple algorithm used to implement the dynamic compensation in feedforward controllers. We showed how to tune the lead/lag, or adjust τ_{ld} and τ_{lg}, based on step-testing the process. This method gives an initial tuning for the algorithm. But what if the step testing cannot be done? How do we go about tuning the algorithm? Following are some tuning guidelines.

- If we need to lag the input signal (slow down the effect of the manipulated variable), set the lead to zero and select a lag.
- If we need to lead the input signal (speed up the effect of the manipulated variable), concentrate on the lead term; however, you must also choose a lag. Obviously, do not use a dead time.
- From the response of the lead/lag algorithm to a step change in input, it is clear that if $\tau_{ld} > \tau_{lg}$, it amplifies the input signal. For noisy signals (e.g., flow) do not use ratios greater than 2.
- Because the dead time just adds to the lag, a negative dead time would effectively decrease the net lag if it could be implemented. Thus we could decrease the lag in the lead/lag unit by the positive dead time. That is,

$$\tau_{lg} \text{ to be used} = \tau_{lg} \text{ calculated} + (t_{oD} - t_{oM})$$

Alternatively, we could increase the lead in the lead/lag unit by the negative of the dead time. That is,

$$\tau_{ld} \text{ to be used} = \tau_{ld} \text{ calculated} - (t_{oD} - t_{oM})$$

- If significant dead time is needed, use a lag, with no lead, and a dead time. It would not make sense to delay the signal and then lead it, even if the transfer functions calls for it.

The third comment also refers to the lead/lag unit, specifically to the location of the unit when multiple disturbances are measured and used in the feedforward controller. If linear compensators are implemented, all that is needed is a single lead/lag unit with adjustable gain for each input. The outputs from the units are then added in the summer, as shown in Fig. 7-4.2. When dynamic compensation is required with nonlinear steady-state compensators, the individual lead/lag units should be installed just after each transmitter, that is, on the inputs to the steady-state compensator. This permits the dynamic compensation for each disturbance to be implemented individually. It would be impossible to provide different dynamic compensations after the measurements are combined in the steady-state compensator.

The fourth comment refers to the steady-state portion of the feedforward controller. This section has shown the development of a linear and a nonlinear compensator. The nonlinear compensator has been shown to provide better performance. Often, it is easy to develop this nonlinear compensator using first principles or a steady-state simulation. If the development of a nonlinear compensator is possible, this is the preferred method. However, if this development is not possible, a linear compensator can be set for each input, and a summer. The decision as to which method to use depends on the process.

The fifth and final comment refers to the comparison of feedforward control to cascade, and ratio control. Feedforward and cascade control take corrective action before the controlled variable deviates from the set point. Feedforward control takes corrective action *before, or at the same time as*, the disturbance enters the process. Cascade control takes corrective action before the primary controlled variable is affected but after the disturbance has entered the process. Figure 7-2.6 shows the implementation of feedforward control only, that is, with no feedback compensation. Interestingly, this scheme is similar to the ratio control scheme shown in Fig. 5-2.2. The ratio control scheme does not have dynamic compensation; however, the ratio unit in Fig. 5-2.2 provides the same function as the gain unit shown in Fig. 5-2.6. Thus, we can say that ratio control is the simplest form of feedforward control.

7-7 ADDITIONAL DESIGN EXAMPLES

Example 7-7.1. An interesting and challenging process is control of the liquid level in a boiler drum. Figure 7-7.1 is a schematic of a boiler drum. The control of the level in the drum is very important. A high level may result in carrying liquid water over into the steam system; a low level may result in tube failure due to overheating for lack of water in the boiling surfaces.

Figure 7-7.1 shows steam bubbles flowing upward through riser tubes into the water; this presents an important phenomenon. The specific volume (volume/mass) of the bubbles is very large, and therefore these bubbles displace the water. This results in a higher apparent level than the level due to liquid water only. The presence of these bubbles also presents a problem under transient conditions. Consider the situation when the pressure in the steam header drops because of an increased demand for steam by the users. This drop in pressure results in a certain quantity of liquid water flashed into steam bubbles. These new bubbles tend to increase the apparent level in the drum. The drop in pressure also causes the volume of the existing bubbles to expand, further increasing the apparent level. This surge in level

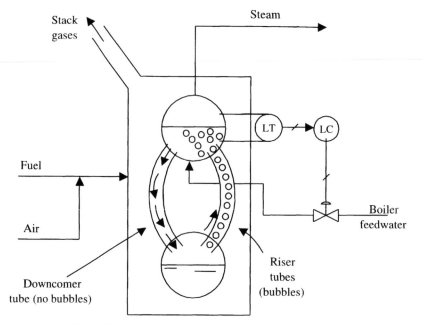

Figure 7-7.1 Single-element control in a boiler drum.

resulting from a decrease in pressure is called *swell*. An increase in steam header pressure caused by a decreased demand for steam by users has the opposite effect on the apparent level and is called *shrink*.

The swell/shrink phenomena combined with the importance of maintaining a good level makes the level control even more critical. The following paragraphs develop some of the level control schemes presently used in practice.

Drum level control is accomplished by manipulating the flow of feedwater. Figure 7-7.1 shows the simplest type of level control, referred to as *single-element control*. A standard differential pressure sensor-transmitter is used. This control scheme relies only on the drum level measurement and therefore must be reliable. Under frequent transients the swell/shrink phenomena do not render a reliable measurement; consequently, a control scheme that compensates for these phenomena is required. A single element is good for boilers that operate at a constant load.

The new control scheme, called *two-element control* and shown in Fig. 7-7.2, is essentially a feedforward/feedback control system. The idea behind this scheme is that the major reason for level changes are changes in steam demand, and that for every pound of steam produced, a pound of feedwater should enter the drum; there should be a mass balance. The output signal from the flow transmitter provides the feedforward part of the scheme, while the level controller provides the feedback compensation for any unmeasured flows, such as blowdown. The feedback controller also helps to compensate for errors in flowmeters.

The two-element control scheme works quite well in many industrial boiler drums. However, there are some systems that exhibit variable pressure drop across the feedwater valve. The two-element control scheme does not compensate directly for this disturbance, and consequently, it upsets drum level control by momentarily

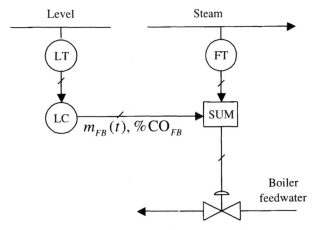

Figure 7-7.2   Two-element control.

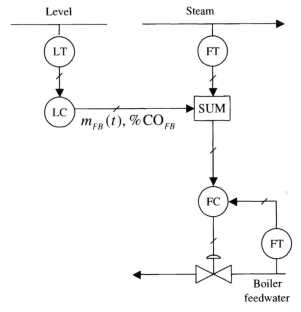

Figure 7-7.3   Three-element control.

upsetting the mass balance. The three-element control scheme, shown in Fig. 7-7.3, provides the required compensation. This scheme provides a tight mass balance during transients. It is interesting to note that all that has been added to the two-element control scheme is a cascade control system.

Example 7-7.2. We now present another industrial example that has proven to be a successful application of feedforward control. The example is concerned with temperature control in the rectifying section of a distillation column. Figure 7-7.4 shows the bottom of the column and the control scheme originally proposed and imple-

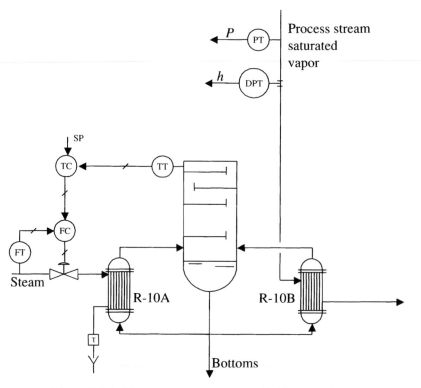

Figure 7-7.4   Temperature control in a distillation column.

mented. This column uses two reboilers. One of the reboilers, R10B, uses a condensing process stream as a heating medium, and the other reboiler, R10A, uses condensing steam. For efficient energy operation, the operating procedure calls for using as much of the process stream as possible. This stream must be condensed anyway, and thus serves as a "free" energy source. Steam flow is used to control the temperature in the column.

After startup of this column, it was noticed that the process stream serving as heating medium experienced changes in flow and in pressure. These changes acted as disturbances to the column and consequently, the temperature controller needed to compensate continually for these disturbances. The time constants and dead time in the column and reboilers complicated the temperature control. After the problem was studied, it was decided to use feedforward control. A pressure transmitter and a differential pressure transmitter had been installed in the process stream, and from them the amount of energy given off by the stream in condensing could be calculated. Using this information the amount of steam required to maintain the temperature at set point could also be calculated, and thus corrective action could be taken before the temperature deviated from the set point. This is a perfect application of feedforward control.

Specifically, the procedure implemented was as follows. Because the process stream is pure and saturated, the density ρ is a function of pressure only. Therefore, using a thermodynamic correlation, the density of the stream can be obtained:

$$\rho = f_1(P) \tag{7-7.1}$$

Using this density and the differential pressure h obtained from the transmitter DPT, the mass flow of the stream can be calculated from the orifice equation:

$$w = K_o \sqrt{h\rho} \tag{7-7.2}$$

where K_o is the orifice coefficient.

Also, knowing the stream pressure and using another thermodynamic relation, the latent heat of condensation λ can be obtained:

$$\lambda = f_2(P) \tag{7-7.3}$$

Finally, multiplying the mass flow rate times the latent heat, the energy q_1 given off by the process stream in condensing is obtained:

$$q_1 = w\lambda \tag{7-7.4}$$

Figure 7-7.5 shows the implementation of Eqs. (7-7.1) through (7-7.4) and the rest of the feedforward scheme. Block PY48A performs Eq. (7-7.1), block PY48B performs Eq. (7-7.2), block PY48C performs Eq. (7-7.3), and block PY48D performs Eq. (7-7.4). Therefore, the output of PY48D is q_1, the energy given off by the condensing process stream.

To complete the control scheme, the output of the temperature controller is considered to be the total energy required q_{total} to maintain the temperature at its set point. Subtracting q_1 from q_{total}, the energy required from the steam, q_{steam}, is determined:

$$q_{steam} = q_{total} - q_1 \tag{7-7.5}$$

Finally, dividing q_{steam} by the latent heat of condensation of the steam, h_{fg}, the required steam flow w_{steam} is obtained:

$$w_{steam} = \frac{q_{steam}}{h_{fg}} \tag{7-7.6}$$

Block TY51 performs Eqs. (7-7.5) and (7-7.6) and its output is the set point to the steam flow controller FC. The latent heat of condensation of steam, h_{fg}, was assumed constant in Eq. (7-7.6). If the steam pressure varies, the designer may want to make h_{fg} a function of this pressure.

Several things must be noted in this feedforward scheme. First, the feedforward controller is not one equation but several. This controller was obtained using several process engineering principles. This makes process control fun, interesting, and challenging. Second, the feedback compensation is an integral part of the control strategy. This compensation is q_{total} or total energy required to maintain temperature set point. Finally, the control scheme shown in Fig. 7-7.5 does not show dynamic compensation. This compensation may be installed later if needed.

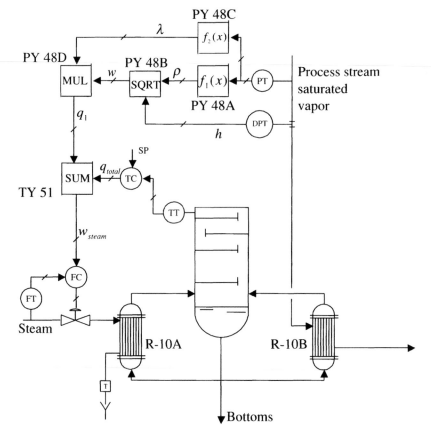

Figure 7-7.5   Implementation of feedforward control.

7-8 SUMMARY

In this chapter we have presented in detail the concept, design, and implementation of feedforward control. The technique has been shown to provide significant improvement over the control performance provided by feedback control. However, undoubtedly the reader has noticed that the design, implementation, and operation of feedforward control requires a significant amount of engineering, extra instrumentation, understanding, and training of the operating personnel. All of this means that feedforward control is more costly than feedback control and thus must be justified. The reader must also understand that feedforward is not the solution to all the control problems. It is another good "tool" to aid feedback control in some cases.

It was shown that feedforward control is generally composed of steady-state compensation and dynamic compensation. Not in every case are both compensations needed. Finally, feedforward control must be accompanied by feedback compensation. It is actually feedforward/feedback that is implemented.

REFERENCES

1. F. G. Shinskey, Feedforward control applied, *ISA Journal*, November 1963.
2. C. A. Smith and A. B. Corripio, *Principles and Practice of Automatic Process Control*, 2nd ed., Wiley, New York, 1997.

PROBLEM

7-1. Problem 5-12 describes a furnace with two sections and a single stack. Referring to that process, if the flow of hydrocarbons changes, the outlet temperature will deviate from the set point, and the feedback controller will have to react to bring the temperature back to the set point. This seems a natural use of feedforward control. Design this strategy for each section.

CHAPTER 8

DEAD-TIME COMPENSATION

It is well established that the presence of dead time in processes adversely affects the stability and therefore the performance of control systems. The longer the dead time, the less aggressive the controller must be tuned to maintain stability. This lack of "aggressivity" in the controller affects the control performance obtained from the strategy.

In this chapter we present a couple of controllers that have been developed in an effort to obtain improved control performance on processes with "significant" dead time. Obviously, even if a process has significant dead time, but the control performance obtained using a simple PID controller is satisfactory, there is no justification for implementation of the technique presented here. The interpretation of when the dead time is significant varies. The ratio t_o/τ is commonly used as a measurement of the effect of dead time. A ratio of zero (as in flow and liquid pressure loops) shows no dead-time effect. Usually, these loops are not difficult to control and they have good performance. As this ratio increases, the dead time becomes more important. Some control experts claim that a ratio of 1.0 indicates a significant effect, while others believe that ratios greater than 1.5 indicate significant effect. Actually, it is up to the control engineer to decide when the presence of dead time is affecting the control performance of his or her process. However, the t_o/τ ratio can provide an indication of when to start looking. The controllers presented in this chapter are referred to as the *Smith predictor* and *Dahlin's controller.*

8-1 SMITH PREDICTOR DEAD-TIME COMPENSATION TECHNIQUE

This section presents the Smith predictor dead-time compensation that was first presented by O. J. M. Smith in 1957 [1]. This significant contribution by Smith was not only the first attempt to design a control strategy to compensate for dead time, but it was also a contribution to what is known today as *model predictive control.* The

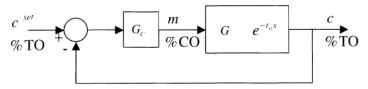

Figure 8-1.1   Block diagram of process.

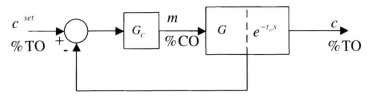

Figure 8-1.2   Block diagram showing Smith's idea.

idea behind this technique is not only very simple to understand, but also very appealing.

Consider Fig. 8-1.1, showing a simple general block diagram. The diagram shows that the process is composed of a transfer function G and a dead time t_o. Since t_o is the source of the problem, it would be great if the controlled variable could be measured before it enters the dead time, as shown in Fig. 8-1.2. However, this is usually not possible because the dead time is not a distinct part of the process, but rather, it is distributed throughout the process.

To get around this problem, Smith proposed to model the process by a first-order-plus-dead time model, that is,

$$Ge^{-t_o s} \approx \frac{Ke^{-t_o s}}{\tau s + 1}$$

The gain and time constant part of this model can then be used to predict the effect of the output signal from the controller; this is shown in Fig. 8-1.3. If this was a perfect model (utopia!), the model would predict the controlled variable before it enters the dead time. Therefore, control action could be taken based on this prediction. However, Smith was realistic and proposed to find the error of the prediction and added to the same prediction as shown in Fig. 8-1.4.

Analyzing Fig. 8-1.4 in detail, it shows that whenever the controller changes its output, in an effort to correct an error, it immediately receives a feedback signal. Branch A provides this immediate response, or "prediction." Branch B provides the error correction continuously. The final effect is that the controller does not feel the effect of the dead time, and thus it can be tuned more aggressively.

As mentioned previously, this strategy was developed in 1957. However, at that time the tools available to implement the dead-time term were not available. That is, with analog instrumentation the implementation of the dead-time term is either impossible or very difficult to accomplish. Computer control systems provide this necessary power.

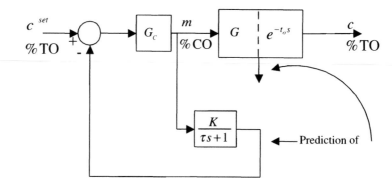

Figure 8-1.3   Smith's idea.

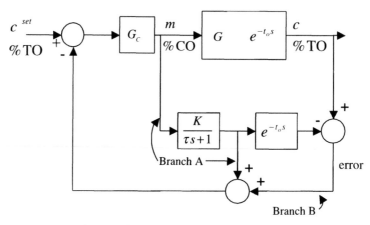

Figure 8-1.4   Smith predictor technique.

8-2   DAHLIN CONTROLLER

Dahlin introduced a method for synthesizing computer feedback controllers [2]. When the process has dead time, the Dahlin synthesis method results in a PID controller with an added term that provides dead-time compensation. In fact, the dead-time compensation term is exactly equivalent to the Smith predictor. The basic advantage of the Dahlin method is that it provides tuning parameters for the PID part of the controller, while the Smith predictor does not.

   A computer controller computes the controller output at regular intervals of time called *sample times*. The period of time between samples is called the *sample time T*. It is convenient to compute the increment in controller output at each sample $\Delta m(k)$ and then add it to the previous controller output $m(k - 1)$ to obtain the updated controller output $m(k)$, where k represents the kth sample. For example, a computer PI controller computes the controller output in the following manner:

$$\Delta m(k) = K_C \left[e(k) - e(k - 1) + \frac{T}{\tau_I} e(k) \right]$$

$$m(k) = m(k - 1) + \Delta m(k)$$

(8-2.1)

where $e(k)$ is the error at the kth sample, K_c the controller gain, T the sample time, and τ_I the integral time.

The Dahlin dead-time compensation controller adds one term to the calculation of the controller output, as follows:

$$m(k) = m(k-1) + \Delta m(k) + (1-q)[m(k-N-1) - m(k-1)] \qquad (8\text{-}2.2)$$

where N is the integer ratio of the dead time to the time constant:

$$N = \text{INT}\left(\frac{t_o}{T}\right) \qquad (8\text{-}2.3)$$

and q is an adjustable parameter in the range of zero to 1.0 which is related to the tuning parameter λ of the controller synthesis method (see Section 3-4.2) as follows:

$$q = e^{-T/\lambda} \qquad (8\text{-}2.4)$$

The last term of Eq. (8-2.2) provides dead-time compensation equivalent to the Smith predictor. Notice that if the dead time is zero, $N = 0$ and the last term of Eq. (8-2.2) vanishes.

The tuning of the controller follows the controller synthesis method of Section 3-4.2. Since the Dahlin controller compensates for the dead time in the process, the controller is tuned as if the process had no dead time; that is, use only the process gain K and time constant τ. The formulas of Section 3-4.2 give us the following results for the Dahlin controller:

$$K_c = \frac{\tau}{K\lambda} \qquad \tau_I = \tau \qquad (8\text{-}2.5)$$

and the derivative time is zero since the process dead time is taken as zero. If we were to use the first guesses of τ from Section 3-4.2, the initial proportional gain would be infinity. This is because, theoretically, if the controller compensates perfectly for dead time, a very high gain would result in an almost perfect control without oscillation. In practice, since the process does not normally match the FOPDT model, a conservative value of the gain should be used. This author recommends a first guess of $\lambda = 0.1\,\tau$.

The following example compares the response of the Dahlin dead-time compensation controller to that of a PID controller.

Example 8-2.1. A step test of the temperature controller of a heat exchanger gives the following FOPDT parameters:

$$K = 1\%\text{TO}/\%\text{CO} \qquad \tau = 0.56\,\text{min} \qquad t_0 = 0.27\,\text{min}$$

A computer-based controller with a sample time $T = 0.05\,\text{min}$ is used to control the temperature. The CSM method of Section 3-4.2 results in the following tuning for a PID controller with $\tau = 0.2(0.27) = 0.054\,\text{min}$:

$$K_C = 1.73\,\%\text{CO}/\%\text{TO}$$
$$\tau_1 = 0.56\text{min}$$
$$\tau_D = 0.13\text{min}$$

The parameters for the Dahlin dead-time compensation controller with $\lambda = 0.056$ are

$$K_C = 10.0\,\%\text{CO}/\%\text{TO}$$
$$\tau_1 = 0.56\text{min}$$
$$\tau_D = 0\text{min}$$
$$N = \text{INT}(0.27/0.05) = 5$$
$$q = e^{-(0.05/0.056)} = 0.41$$

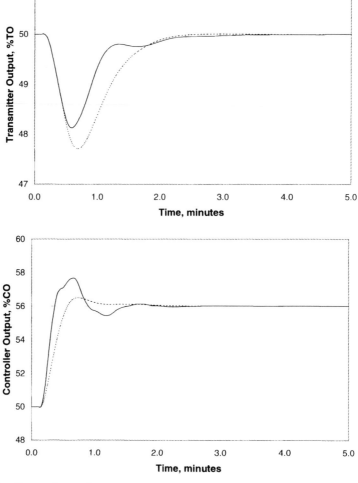

Figure 8-2.1 Comparison of responses to a disturbance input: PI with dead-time compensation (solid line) versus standard PID (dashed line).

Figure 8-2.1 compares the responses of the two controllers to a step change in process flow to the exchanger. The PI controller with dead-time compensation does slightly well than the normal PID controller by keeping the deviation from set point smaller. This better performance is caused by the higher controller gain afforded by dead-time compensation. The higher gain also results in a higher overshoot in the response of the controller output.

8-3 SUMMARY

In this brief chapter we have presented two controllers that may provide improved control performance in processes with significant dead times. These controllers were developed many years ago. Today's DCSs and other available process computers make implementation of these controllers very realistic.

REFERENCES

1. O. J. M. Smith, Closer control of loops with dead time, *Chemical Engineering Progress*, 53:217–219 May 1957.
2. E. B. Dahlin, Designing and tuning digital controllers, *Instruments and Control Systems*, 41:77 June 1968.

CHAPTER 9

MULTIVARIABLE PROCESS CONTROL

Up to this point in our study of automatic process control only processes with a single controlled variable and manipulated variable have been considered. These processes are often referred to as *single-input, single-output* (SISO) *processes*. Frequently, however, processes with more than one input and output variables are encountered; these are named *multivariable processes* or *multiple-input, multiple-output* (MIMO) *processes*. Some examples are shown in Fig. 9-1.1.

Figure 9-1.1a depicts a blending tank where two streams are mixed. Both streams are composed of water and salt; stream 1 is more concentrated in salt. In this process it is necessary to control the outlet flow and outlet mass fraction of component salt. To accomplish this control objective, the valves that regulate the flows of streams 1, W_1, and 2, W_2, are used. Figure 9-1.1b shows a chemical reactor for which it is necessary to control the outlet temperature and composition. The manipulated variables in this case are the cooling water flow and the outlet flow. Figure 9-1.1c shows an evaporator with the level and outlet concentration as controlled variables and with the outlet process flow and steam flow as the manipulated variables. Finally, Fig. 9-1.1d depicts a typical distillation column with five controlled variables: column pressure, distillate composition, accumulator level, base level, and tray temperature. To accomplish this control, five manipulated variables are used: cooling water flow to the condenser, distillate flow, reflux flow, bottoms flow, and steam flow to the reboiler.

The examples above show that the control of these processes can be quite complex and challenging to the operation. There are usually four questions that the engineer must ask when faced with a control problem of this type:

1. Which is the best pairing of controlled and manipulated variables?
2. How does the interaction affect the stability of the loops?
3. How should the feedback controllers be tuned in a multivariable scheme?

4. Can something be done with the control scheme to break, or minimize, the interaction between loops?

In this chapter we show how to answer theses questions using simple, proven techniques.

9-1 PAIRING CONTROLLED AND MANIPULATED VARIABLES

The first question posed relates to how to pair controlled and manipulated variables. Often this decision is simple, but at other times it is not as simple. Examples of difficult cases are the blending system shown in Fig. 9-1.1a and the chemical reactor shown in Fig. 9-1.1b. For these cases there is a technique that has proven to be successful in numerous processes. A 2×2 process, shown in Fig. 9-1.2, is used to present the technique. Once this is done, we extend the technique to an $n \times n$ process. (In

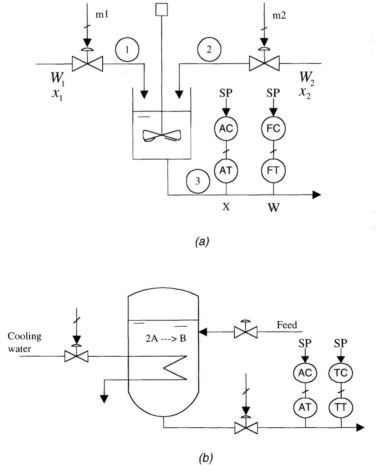

(a)

(b)

Figure 9-1.1 Examples of multivariable control systems: (a) blending tank; (b) chemical reactor; (c) evaporator; (d) distillation column.

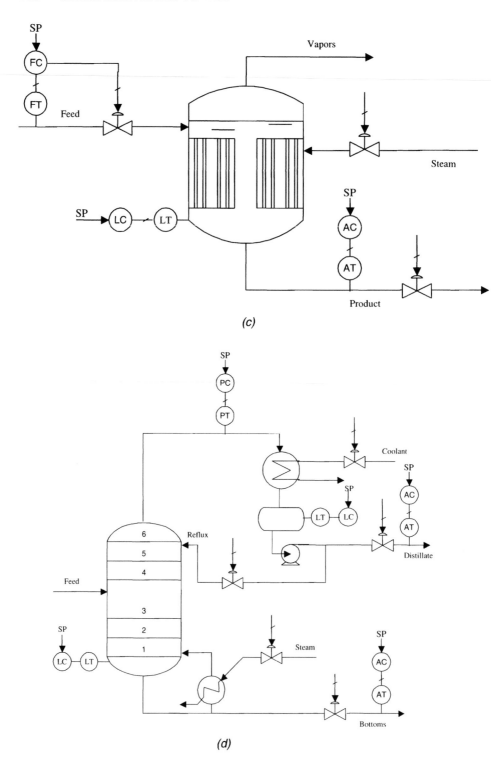

(c)

(d)

Figure 9-1.1 *Continued.*

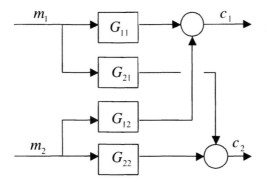

Figure 9-1.2   Block diagram of a 2 × 2 multivariable process.

$$
\begin{array}{c c c}
 & m_1 & m_2 \\
c_1 & K_{11} & K_{12} \\
\\
c_2 & K_{21} & K_{22}
\end{array}
$$

Figure 9-1.3   Steady-state gain matrix.

this notation the first n is the number of controlled variables and the second n is the number of manipulated variables.)

If we don't know how to decide but a decision has to be made, it makes sense to control each controlled variable with the manipulated variable that has the greatest influence on it. In this context, influence and process gain have the same meaning; consequently, to make a decision we must find all process gains (four gains for a 2 × 2 system) of the process. The following are the open-loop process gains of interest:

$$
K_{11} = \left.\frac{\Delta c_1}{\Delta m_1}\right|_{m_2} \qquad K_{12} = \left.\frac{\Delta c_1}{\Delta m_2}\right|_{m_1}
$$

$$
K_{21} = \left.\frac{\Delta c_2}{\Delta m_1}\right|_{m_2} \qquad K_{22} = \left.\frac{\Delta c_2}{\Delta m_2}\right|_{m_1}
$$

where the notation K_{ij} refers to the gain relating the ith controlled variable to the jth manipulated variable.

The four gains can be arranged in the form of a matrix to give a more graphical/mathematical description of their relationship to the controlled and manipulated variables. This matrix is called the *steady-state gain matrix* (SSGM) and is shown in Fig. 9-1.3.

From this SSGM the combination of the controlled and manipulated variables that yields the largest absolute number in each row may appear to be the one that should be chosen. For example, if $|K_{12}|$ is larger than $|K_{11}|$, m_2 is chosen to control

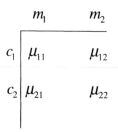

Figure 9-1.4   Relative gain matrix.

c_1. However, this method of choosing the pairing of controlled and manipulated variables is not correct because it suffers three weaknesses: (1) the comparison of the second row may yield the use of the same manipulated variable; (2) under closed-loop operation the gains may vary; and more important, (3) the gains may have different units. Thus it is not a fair comparison; the matrix, as it stands, is dependent on units.

A technique developed by Bristol [1] has been proposed to normalize the terms in the matrix, making them independent of the units, taking into consideration the closed-loop gains, and ensuring that no manipulated variable is chosen more than once. This technique, called the *relative gain analysis* or interaction measure, yields the relative gain matrix (RGM), which is then used to reach a decision. The RGM is shown in Fig. 9-1.4.

The *relative gain terms* in the RGM are defined as follows:

$$\mu_{12} = \frac{\partial c_1 / \partial m_2 |_{m_1}}{\partial c_1 / \partial m_2 |_{c_2}} \approx \frac{\Delta c_1 / \Delta m_2 |_{m_1}}{\Delta c_1 / \Delta m_2 |_{c_2}} = \frac{K_{12}}{K'_{12}} \tag{9.1-1}$$

or, in general,

$$\mu_{ij} = \frac{\partial c_i / \partial m_j |_m}{\partial c_i / \partial m_j |_c} = \frac{K_{ij}}{K'_{ij}} \tag{9.1-2}$$

Let us make sure that we understand the meaning and significance of all the terms in Eq. (9-1.2). The numerator

$$\left. \frac{\partial c_i}{\partial m_j} \right|_m \approx \left. \frac{\Delta c_i}{\Delta m_j} \right|_m$$

is the open-loop steady-state gain K_{ij}, defined previously. That is, this is the gain of m_j on c_i when all other manipulated variables are kept constant. The denominator

$$\left. \frac{\partial c_i}{\partial m_j} \right|_c \approx \left. \frac{\Delta c_i}{\Delta m_j} \right|_c$$

is the closed-loop gain, K'_{ij}. That is, this is the gain of m_j on c_i when all other loops are closed and all controllers have integral action, thus returning the controlled variables to their corresponding set points. Therefore, we can write

$$\mu_{ij} = \frac{\text{gain when all other loops are open}}{\text{gain when all other loops are closed}} = \frac{K_{ij}}{K'_{ij}}$$

As seen from the definition of μ_{ij}, this term takes into consideration the gain under closed-loop conditions and is a dimensionless number. In addition, a property of the RGM is that the summation of all the terms in each row and in each column must equal 1. (This means that for a 2×2 system only one term has to be evaluated, and the others can be obtained using this property. For a 3×3 system, which has nine terms in the matrix, only four independent terms have to be evaluated, and the other five can be obtained using this property.) Thus the RGM resolves all three weaknesses of the SSGM, and therefore it can be used to decide how to pair controlled and manipulated variables.

A complete understanding of the meaning/significance of μ_{ij} is important before proceeding. From the definition of μ_{ij}, notice that it is essentially a measure of the effect of closing all other loops on the process gain for a given controlled and manipulated variable pair. That is, if $\mu_{12} = \frac{4}{5} = 0.8$, it means that when the other loops are closed, the "effect" of a change in m_2 on c_1 is larger than when the other loops are open. Specifically, the value says that the gain when the other loops are open is only 80% of the gain when the other loops are closed. Thus the numerical value of μ_{ij} is a measure of the interaction between the control loops.

If $\mu_{ij} = 1$, the process gain is the same with all other loops open or closed; of course, this is good! This indicates either no interaction between the particular loop and all other loops, or possible offsetting interactions. The greater the deviation from 1, the greater the loop interaction.

If $\mu_{ij} \approx 0$, it may be due to one of two possibilities. One is that the open-loop gain, $\partial c_i / \partial m_j|_m$, is either zero or very small. In this case m_j does not affect c_i, or hardly any, when all other loops are open. Alternatively, the closed loop is so large that $\mu_{ij} \approx 0$. This means that to keep the other controlled variables constant, the other loops interact significantly with the loop in question. In either case, this is no good! Either of the two possibilities indicates that c_i should not be controlled manipulating m_j.

If $\mu_{ij} \approx \infty$ (very large), it may be due to one of two possibilities. One, the closed-loop gain, $\partial c_i / \partial m_j|_c$, is either zero or very small. This means that when the other loops are in automatic mode, the loop in question cannot be controlled because m_j does not affect c_i, or hardly any. Alternatively, the open-loop gain is very large. This means that when the other loops are in manual mode, the effect of m_j on c_i is very large, whereas it is not the case when the others are in automatic. Obviously, this condition is also no good!

The preceding discussion has illustrated the significance of μ_{ij}. In general, values of μ_{ij} close to 1 represent controllable combinations of controlled and manipulated variables. Values of μ_{ij} approaching values of zero or infinity represent uncontrollable combinations.

With this background we can understand the *pairing rule* first presented by Bristol [1]: *To minimize the interaction between loops, always pair on RGM elements that are closest to 1.0. Avoid negative pairings.* The proposed pairing rule is easy and convenient to use. Realize that only steady-state information is needed. This is certainly an advantage since this information can even be found during the process design stage. Thus, it does not require the process to be in operation. In the next section we discuss in more detail the calculation of these gains.

To close this presentation let us look at two possible RGMs to further understand what the μ_{ij} terms are telling us about the control system. Consider the following RGM:

$$
\begin{array}{c|cc}
 & m_1 & m_2 \\
\hline
c_1 & 0.2 & 0.8 \\
 & & \\
c_2 & 0.8 & 0.2 \\
\end{array}
$$

The terms $\mu_{11} = \mu_{22} = 0.2 = \frac{1}{5}$ indicate that for this pairing the gain of each loop increases by a factor of 5 when the other loop is closed. The terms $\mu_{12} = \mu_{21} = 0.8 = \frac{4}{5}$ indicate that for this pairing the gain increases only by a factor of 1.25. This explains why the $c_1 - m_2$ and $c_2 - m_1$ pairing is the correct one.

Consider another RGM:

$$
\begin{array}{c|cc}
 & m_1 & m_2 \\
\hline
c_1 & 2 & -1 \\
 & & \\
c_2 & -1 & 2 \\
\end{array}
$$

The terms $\mu_{11} = \mu_{22} = 2 = 1/0.5$ indicate that the gain of each loop is cut in half when the other loop is closed. The terms $\mu_{12} = \mu_{21} = -1.0$ indicate that the gain of each loop changes sign when the other loop is closed. Certainly, this is undesirable because it means that the action of the controller depends on whether the other loop is closed or open. This explains why the correct pairing is $c_1 - m_1$ and $c_2 - m_2$.

9-1.1   Obtaining Process Gains and Relative Gains

The first information needed to obtain the relative gain terms is the steady-state open-loop gains, K_{ij}. There are three different ways to calculate these gains:

1. Using the step test method learned in Chapter 2. This is the method used to obtain the information to tune feedback and cascade controllers and to design feedforward controllers. Thus we are quite familiar with it.
2. Starting analytically from the equations that describe the process.
3. By the use of a flowsheet simulator.

To obtain these steady-state open-loop gains analytically, the equations that describe the process are written first. From these equations the gains are then evaluated. Using the blending process of Fig. 9-1.1a as an example, the outlet flow W and the outlet mass fraction of salt x are to be controlled. Because there are two components, salt and water, two independent mass balances can be written. A steady-state total mass balance provides the first equation,

$$W_1 + W_2 = W \tag{9-1.3}$$

A mass balance on salt provides the other equation,

$$W_1 x_1 + W_2 x_2 = Wx$$

or

$$x = \frac{W_1 x_1 + W_2 x_2}{W}$$

and substituting Eq. (9-1.3) into the above,

$$x = \frac{W_1 x_1 + W_2 x_2}{W_1 + W_2} \qquad (9\text{-}1.4)$$

In this 2×2 system there are four gains of interest: K_{W1}, K_{W2}, K_{x1}, and K_{x2}. From Eq. (9-1.3) the first two gains can be evaluated:

$$K_{W1} = \left.\frac{\partial W}{\partial W_1}\right|_{W_2} = 1 \quad \text{and} \quad K_{W2} = \left.\frac{\partial W}{\partial W_2}\right|_{W_1} = 1$$

From Eq. (9-1.4) the other two gains are evaluated:

$$K_{x1} = \left.\frac{\partial x}{\partial W_1}\right|_{W_2} = \frac{W_2(x_2 - x_1)}{(W_1 + W_2)^2} \quad \text{and} \quad K_{x2} = \left.\frac{\partial x}{\partial W_2}\right|_{W_1} = \frac{W_1(x_1 - x_2)}{(W_1 + W_2)^2}$$

The steady-state gain matrix is then written as

	W_1	W_2
W	1	1
x	$\dfrac{W_2(x_2 - x_1)}{(W_1 + W_2)^2}$	$\dfrac{W_1(x_1 - x_2)}{(W_1 + W_2)^2}$

For this blending process, development of the set of describing equations and evaluation of the gains were fairly simple. For some processes these are not easily done; examples are the chemical reactor and the distillation column, shown in Fig. 9-1.1. Fortunately, however, the design of most processes is usually done with the use of flowsheet simulators, such as ASPEN, HYSIM, CHEMCAD, and ProII. From these simulators it is usually simple to evaluate the required gains. For a 2×2 system three computer runs suffice to obtain the four gains. In this case the following approximation is used: $K_{ij} \approx \left.\Delta c_i / \Delta m_j\right|_m$.

Once the open-loop gains K_{ij} have been obtained, evaluation of the closed-loop gains K'_{ij} and the relative gain terms μ_{ij} is fairly straightforward. For the closed-loop gain there is no need to actually go to the process and evaluate it. We show next how to obtain this closed-loop gain and the relative gain for a 2×2 process; the method is then extended to any higher-order process.

Consider the block diagram for a 2×2 process shown in Fig. 9-1.2. The effect of a change in both manipulated variables on c_1 is expressed as follows:

$$\Delta c_1 = K_{11}\Delta m_1 + K_{12}\Delta m_2 \tag{9-1.5}$$

Similarly, on c_2 we have

$$\Delta c_2 = K_{21}\Delta m_1 + K_{22}\Delta m_2 \tag{9-1.6}$$

To obtain the gain $\partial c_1/\partial m_1|_{c_2} \approx \Delta c_1/\Delta m_1|_{c_2}$, Δc_2 in Eq. (9-1.6) is set to zero:

$$0 = K_{21}\Delta m_1 + K_{22}\Delta m_2$$

or

$$\Delta m_2 = -\frac{K_{21}}{K_{22}}\Delta m_1$$

Substituting this expression for Δm_2 in Eq. (9-1.5) yields

$$\Delta c_1 = K_{11}\Delta m_1 - \frac{K_{12}K_{21}}{K_{22}}\Delta m_1$$

and finally, we obtain

$$K_{11}' = \frac{\Delta c_1}{\Delta m_1}\bigg|_{c_2} = \frac{K_{11}K_{22} - K_{12}K_{21}}{K_{22}} \tag{9-1.7}$$

Note that this closed-loop gain can be evaluated simply by a combination of open-loop gains. There is no need to close any loop nor to have the process operating. Then

$$\mu_{11} = \frac{K_{11}}{K_{11}'} = \frac{K_{11}K_{22}}{K_{11}K_{22} - K_{12}K_{21}} \tag{9-1.8}$$

A similar procedure on each of the other combinations yields

$$\mu_{12} = \frac{K_{12}}{K_{12}'} = \frac{K_{12}K_{21}}{K_{12}K_{21} - K_{11}K_{22}} \tag{9-1.9}$$

$$\mu_{21} = \frac{K_{21}}{K_{21}'} = \frac{K_{12}K_{21}}{K_{12}K_{21} - K_{11}K_{22}} \tag{9-1.10}$$

$$\mu_{22} = \frac{K_{22}}{K_{22}'} = \frac{K_{11}K_{22}}{K_{11}K_{22} - K_{12}K_{21}} \tag{9-1.11}$$

and the relative gain matrix is

$$
\begin{array}{c|cc}
 & m_1 & m_2 \\
\hline
c_1 & \dfrac{K_{11}K_{22}}{K_{11}K_{22} - K_{12}K_{21}} & \dfrac{K_{12}K_{21}}{K_{12}K_{21} - K_{11}K_{22}} \\[3ex]
c_1 & \dfrac{K_{12}K_{21}}{K_{12}K_{21} - K_{11}K_{221}} & \dfrac{K_{11}K_{22}}{K_{11}K_{22} - K_{12}K_{21}}
\end{array}
$$

From this matrix, using the pairing rule presented earlier, the correct combination of controlled and manipulated variables is chosen. It is easily shown from the matrix that the terms in each row and each column add up to 1. The dimensionality consistency of each term is also easily shown.

Applying the relative gain matrix to the blending process yields

$$
\begin{array}{c|cc}
 & W_1 & W_2 \\
\hline
W & \dfrac{W_1}{W} & \dfrac{W_2}{W} \\[3ex]
x & \dfrac{W_2}{W} & \dfrac{W_1}{W}
\end{array}
$$

If $W_1/W > 0.5$, the correct pairing is $W - W_1$ and $x - W_2$. If $W_1/W < 0.5$, the correct pairing is $W - W_2$ and $x - W_1$. A value of $W_1/W = 0.5$ yields all μ_{ij}'s equal to 0.5. This value for a 2×2 system indicates the highest degree of interaction when the interaction is positive (presented next).

The steps taken to develop the RGM for a 2×2 system required only simple algebra. For a higher-order system the same procedure could be followed; however, more algebraic steps must be taken to reach a final solution. For these higher-order systems matrix algebra can be used to simplify the development of the RGM. The procedure, as proposed by Bristol [1], is as follows: *Calculate the transpose of the inverse of the steady-state gain matrix and multiply each term of the new matrix by the corresponding term of the original matrix. The terms thus obtained are the terms of the relative gain matrix.*

This procedure might look out of reach for those unfamiliar with matrix algebra. But given the utility of this method, it is worthwhile to surpass this difficulty. There are digital computer programs, and even handheld calculators, that can easily crank out the necessary numbers.

9-1.2 Positive and Negative Interactions

Positive interaction is the interaction experienced when all the relative gain terms are positive. It is interesting to see under what conditions this type of interaction results. Much can be learned from the expression for μ_{11} in a 2×2 system

$$
\mu_{11} = \frac{K_{11}K_{22}}{K_{11}K_{22} - K_{12}K_{21}} = \frac{1}{1 - K_{12}K_{21}/K_{11}K_{22}} \tag{9-1.12}
$$

If there is an odd number of positive K_{ij}'s, the value of μ_{11} will be positive, and furthermore, its numerical value will be between 0 and 1.

Positive interaction is the most common type of interaction in multivariable control systems. In these systems each loop helps the other. To understand what we mean by this, consider the blending control system shown in Fig. 9-1.5, and its block diagram shown in Fig. 9-1.6. For this process the gains of the control valves are positive since both valves are fail-closed. The gains K_{11}, K_{21}, and K_{12} are also positive, while the gain K_{22} is negative. The flow controller G_{C1} is reverse acting, while the analyzer controller G_{C2} is direct acting. Assume now that the set point to the flow controller decreases; the flow controller, in turn, decreases its output to close the valve to satisfy its new set point. This will cause the outputs from valve

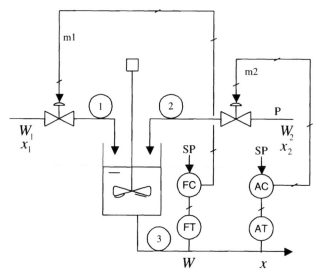

Figure 9-1.5   Control system for blending process.

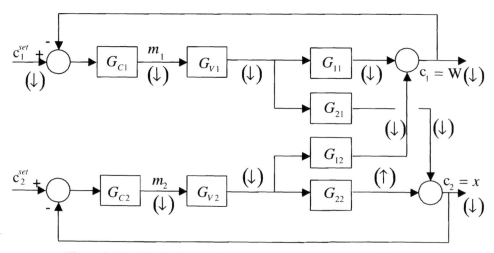

Figure 9-1.6   Block diagram showing how signals and variables move.

G_{V1} and process G_{11} to decrease. Because K_{21} is positive, the output from G_{21} also decreases, resulting in lowering the analysis, x. When this happens, the analysis controller G_{C2} also decreases its output. This causes the outputs from G_{V2} and G_{12} to decrease and the output from G_{22} to increase. Figure 9-1.6 shows the arrows that indicate the directions that each output moves. The figure clearly shows that the outputs from G_{11} and G_{12} both decrease. This is what we mean by "both loops help each other."

When there are an even number of positive values of K_{ij}'s, or an equal number of positive and negative values of K_{ij}'s, the value of μ_{ij}'s will either be less than 0 or greater than 1. In either case there will be some μ_{ij} with negative values in the same row and column. The interaction in this case is said to be a negative interaction. It is important to realize that for a relative gain term to be negative, the signs of the open- and closed-loop gains must be different. This means that the action of the controller must change when the other loops are closed. For this type of interaction the loops "fight" each other.

9-2 INTERACTION AND STABILITY

The second question posed at the beginning of the chapter is related to the effect of the interaction on the stability of multivariable control systems. We first address this question to a 2×2 system; consider Fig. 9-1.6.

As explained in Chapter 7, the roots of the characteristic equation define the stability of control loops. For the system of Fig. 9-1.6 the characteristic equations for loop 1 by itself (when loop 2 is in manual) is

$$1 + G_{C1}G_{V1}G_{11} = 0 \tag{9-2.1}$$

and equally for loop 2,

$$1 + G_{C2}G_{V2}G_{22} = 0 \tag{9-2.2}$$

The control loops are stable if the roots of the characteristic equation have negative real real parts. To analyze the stability of the *complete* system shown in Fig. 9-1.6, the characteristic equation for the complete system must be determined. This is done using signal flow graphs [2], which yields

$$(1 + G_{C1}G_{V1}G_{11})(1 + G_{C2}G_{V2}G_{22}) - G_{C1}G_{V1}G_{C2}G_{V2}G_{12}G_{21} = 0 \tag{9-2.3}$$

The terms in parentheses are the characteristic equations for the individual loops. Analyzing Eq. (9-2.3), the following conclusions for a 2×2 system can be reached:

1. The roots of the characteristic equation for each individual loop are *not* the roots of the characteristic equation for the complete system. Therefore, it is possible for the complete system to be unstable even though each loop is stable. *Complete system* refers to the condition when both controllers are in automatic mode.

2. For interaction to affect the stability, it must work both ways. That is, both G_{12} and G_{21} must exist; otherwise, the last term in Eq. (9-2.3) disappears, and if each loop is stable, the complete loop is also stable. When the interaction works both ways, the system is said to be *fully coupled* or interactive; otherwise, the system is *partially coupled*. Interaction is not a problem in partially coupled 2×2 systems.

3. The interaction effect on one loop can be eliminated by interrupting the other loop; this is easily done by switching the controller to manual. Suppose that controller 2 is switched to manual; this has the effect of setting $G_{C2} = 0$, leaving the characteristic equation as

$$1 + G_{C1}G_{V1}G_{11} = 0$$

which is the same as if only one loop existed. This may be one reason why many controllers in practice are in manual. Manual changes in the output of controller 2 simply become disturbances to loop 1. Usually, however, it is not necessary to be this drastic to yield a stable system. By simply lowering the gain and/or increasing the reset time of the controller, that is, detuning the controller somewhat, the same effect can be accomplished while retaining both controllers in automatic. The effect of doing this is to move all the roots of Eq. (9-2.3) to the negative side of the real axis.

The preceding paragraphs have described how the interaction on a 2×2 system affects the stability of the system, which is probably the most common multivariable control system. For higher-order systems the same procedure must be followed. However, the conclusions are not as simple to generalize.

9-3   TUNING FEEDBACK CONTROLLERS FOR INTERACTING SYSTEMS

The third question asked at the beginning of the chapter refers to tuning of the feedback controllers in a multivariable environment. The interaction among loops makes the tuning of feedback controllers more difficult. The following paragraphs present some procedures for this tuning; for a more complete discussion, see Shinskey [3] and Smith and Corripio [2]. We first discuss tuning for a 2×2 system and then discuss $n \times n$ systems.

The first step, after proper pairing, is to determine the relative speed of the loops. Then:

1. If one loop is much faster than the other one (say, the dominant time constant, or the time constant of the first-order-plus-dead time approximation, is five times smaller), the fast loop is tuned first, with the other loop in manual. Then the slow loop is tuned with the faster loop in automatic. The tuning procedure and formulas are the same as the procedure and formulas described in Chapters 2 and 3.

2. If both loops are about the same speed of response, and one variable is more important to control than the other one, detune the less important loop by

setting a small gain and a long reset time. This will reduce the effect of the less important loop on the response of the most important loop because the detuned loop will appear to be open.

3. If both loops are about the same speed of response and both variables are of the same importance, each controller should be tuned with the other loop in manual. Then the effect of interaction should be used to adjust the tuning.

   (a) If the interaction is positive, the following is proposed:

$$K'_{Ci} = K_{Ci}\mu_{ii} \tag{9-3.1}$$

   (b) If the interaction is negative, the adjustment must be done by trial and error after both loops are closed.

There is still another procedure, developed by Medina [4], that has proven to work quite well and it is easy to apply. The procedure requires that we know the first-order-plus-dead time approximation to each of the four transfer functions that

---

TABLE 9-3.1   Tuning a 2 × 2 Multivariable Decentralized Feedback Controller

Loop 1 is the loop with the smallest $(t_o/\tau)_{ii}$ ratio. Loop 2 is the one with the largest $(t_o/\tau)_{ii}$ ratio. The formulas presented here are to tune loop 2. Loop 1 is tuned by the user by whatever method he or she desires.

PI–PI Combination

$$K_{C2} = \frac{\tau_{22}}{K_{22}}A; \qquad \tau_{I2} = \tau_{22}$$

$$A = \frac{1}{1-\gamma}\frac{1}{\lambda+t_{o22}}; \qquad \gamma = \frac{K_{12}K_{21}}{K_{11}K_{22}}$$

$$\ln\frac{\lambda}{t_{o22}} = 1.104\gamma + 1.124\gamma^2 + 0.066\left(\frac{t_{o12}t_{o21}}{t_{o11}t_{o22}}\right) + 0.368\left(\frac{t_{o11}}{t_{o22}}\right) - 0.237\left(\frac{t_{o21}}{t_{o22}}\right) - 0.12\left(\frac{\tau_{12}\tau_{21}}{\tau_{11}\tau_{22}}\right)$$

Formulas are good for $\gamma \le 0.8$.

PI–PID Combination

Loop 1 is PI and loop 2 is PID.

$$K_{C2} = \frac{\tau_{22}}{K_{22}}A; \qquad \tau_{I2} = \tau_{22}; \qquad \tau_{D2} = \frac{t_{o12} + t_{o21} - t_{o11}}{2}$$

$$A = \frac{1}{1-\gamma}\frac{1}{\lambda+t_{o22}}; \qquad \gamma = \frac{K_{12}K_{21}}{K_{11}K_{22}}$$

$$\ln\frac{\lambda}{t_{o22}} = 1.283\gamma + 1.014\gamma^2 + 0.0675\left(\frac{t_{o12}t_{o21}}{t_{o11}t_{o22}}\right) + 0.463\left(\frac{t_{o11}}{t_{o22}}\right) - 0.319\left(\frac{\tau_{11}}{\tau_{22}}\right) - 0.771\left(\frac{\tau_{21}}{\tau_{22}}\right)$$

Formulas are good for $\gamma \le 0.8$.

is, K_{11}, τ_{11}, t_{o11}, K_{12}, τ_{12}, t_{o12}, K_{21}, τ_{21}, t_{o21}, and K_{22}, τ_{22}, t_{o22}. Remember that all the gains must be in %TO/%CO, as used in Chapter 3 to tune controllers. Table 9-3.1 shows the formulas to use.

9-4   DECOUPLING

Finally, there is still one more question to answer: Can something be done with the control scheme to break, or minimize, the interaction between loops? That is, can a control system be designed to decouple the interacting, or coupled, loops? Decoupling can be a profitable, realistic possibility when applied carefully. The relative gain matrix provides an indication of when decoupling could be beneficial. If for the best pairing option, one or more of the relative gains is far from unity, decoupling may help. For existing systems, operating experience is usually enough to decide. There are two ways to design decouplers: from block diagrams or from basic principles.

9-4.1   Decoupler Design from Block Diagrams

Consider the block diagram shown in Fig. 9-4.1. The figure shows graphically the interaction between the two loops. To circumvent this interaction, a decoupler may be designed and installed as shown in Fig. 9-4.2. The decoupler, terms D_{21} and D_{12}, should be designed to cancel the effects of the cross blocks, G_{21} and G_{12}, so that each controlled variable is not affected by the manipulated variable of the other loop. In other words, decoupler D_{21} cancels the effect of manipulated variable m_1 on controlled variable c_2, and D_{12} cancels the effect of m_2 on c_1. In mathematical terms, we design D_{21} so that

$$\left.\frac{\Delta c_2}{\Delta m_1}\right|_{m_2} = 0$$

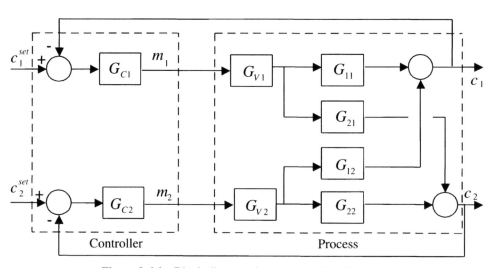

Figure 9-4.1   Block diagram for a general 2×2 system.

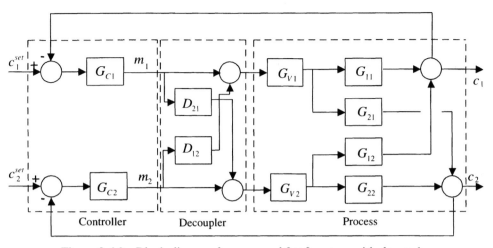

Figure 9-4.2 Block diagram for a general 2×2 system with decoupler.

and D_{12} so that

$$\left.\frac{\Delta c_1}{\Delta m_2}\right|_{m_1} = 0$$

From block diagram algebra,

$$\Delta c_1 = D_{12}G_{V1}G_{11}\Delta m_2 + G_{V2}G_{12}\Delta m_2 \tag{9-4.1}$$

$$\Delta c_2 = D_{21}G_{V2}G_{22}\Delta m_1 + G_{V1}G_{21}\Delta m_1 \tag{9-4.2}$$

Setting $\Delta c_1 = 0$ in Eq. (9-4.1),

$$D_{12} = -\frac{G_{V2}G_{12}}{G_{V1}G_{11}} \tag{9-4.3}$$

and setting $\Delta c_2 = 0$ in Eq. (9-4.2),

$$D_{21} = -\frac{G_{V1}G_{21}}{G_{V2}G_{22}} \tag{9-4.4}$$

Usually, we lump the valve transfer functions with the process unit itself; therefore,

$$D_{12} = -\frac{G_{P12}}{G_{P11}} \tag{9-4.5}$$

$$D_{21} = -\frac{G_{P21}}{G_{P22}} \tag{9-4.6}$$

where $G_{Pij} = G_{Vj}G_{ij}$.

There are several things that should be pointed out. If one looks at the method to design the decoupler, and at its objective, one is reminded of the feedforward controllers. The disturbance to a loop is the manipulated variable of the other loop. Remembering that each process transfer function contains a K_{ij}, a τ_{ij}, and a $t_{o_{ij}}$, decoupler D_{21} looks as follows:

$$D_{21} = -\frac{G_{P21}}{G_{P22}} = -\frac{K_{21}}{K_{22}} \frac{\tau_{22}s + 1}{\tau_{21}s + 1} e^{-(t_{o21} - t_{o22})s}$$

Thus, similar to feedforward controllers, the decoupler is composed of steady-state and dynamic compensations. The difference is that, unlike feedforward controllers, decouplers form part of the feedback loops and therefore they affect the stability. Because of this, the decouplers must be selected and designed with great care.

For more extensive discussion on decoupling, such as partial or steady-state decoupling and decoupling for $n \times n$ systems, the reader is referred to Smith and Corripio [2].

9-4.2 Decoupler Design from Basic Principles

In Section 9-4.1 we showed how to design decouplers using block diagram algebra; thus the decouplers obtained are *linear decouplers*. In this section we present the development of a steady-state decoupler from basic engineering principles. The resulting algorithm is a nonlinear decoupler. The procedure is similar to the one presented in Chapter 7 for designing feedforward controllers.

Consider the blending tank shown in Fig. 9-1.5. In this process there are two components, salt and water; thus two independent mass balances are possible. We start with a total mass balance:

$$W = W_1 + W_2 \tag{9-4.7}$$

A mass balance on salt is used next:

$$W_1 x_1 + W_2 x_2 - Wx = 0 \tag{9-4.8}$$

From Eq. (9-4.7)

$$W_1 = W - W_2 \tag{9-4.9}$$

From Eq. (9-4.8) and using Eq. (9-4.9) yields

$$W_2 = W_1 \frac{x - x_1}{x_2 - x} \tag{9-4.10}$$

Realize that Eqs. (9-4.9) and (9-4.10) provides the manipulated variables W_1 and W_2. However, we have two equations, Eqs. (9-4.9) and (9-4.10), and four unknowns, W_1, W_2, W, and $(x - x_1)/(x_2 - x)$. Thus there are two degrees of freedom. Well, we have two controllers, and we can let the controllers provide two of the unknowns.

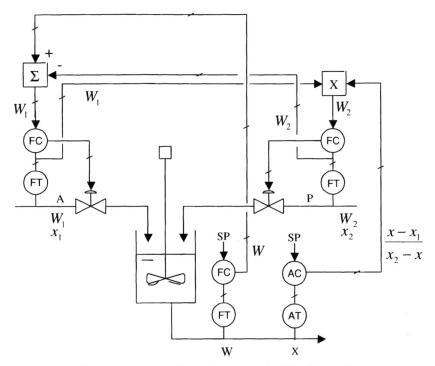

Figure 9-4.3   Nonlinear decoupler for blending tank.

For example, we can call the output of the flow controller W, and we can call the output of the analyzer controller $(x - x_1)/(x_2 - x)$. Figure 9-4.3 shows the control scheme. The decoupler shown provides only steady-state compensation. The information for this compensation is usually the easiest to obtain. To provide dynamic compensation, lead/lag, and/or dead time, dynamic data are usually required. The discussion of the various compensations on feedforward is also very applicable to this chapter.

9-5   SUMMARY

In this chapter we have presented an introduction to the most important aspects of multivariable control. Decentralized controllers, simple feedback controllers, were used. We did not present the subject of multivariable controllers such as dynamic matrix control (DMC).

REFERENCES

1. E. H. Bristol, On a new measure of interaction for multivariable process control, *Transactions IEEE*, January 1966.
2. C. A. Smith and A. B. Corripio, *Principles and Practice of Automatic Process Control*, 2nd ed., Wiley, New York, 1997.
3. F. G. Shinskey, *Process Control Systems*, McGraw-Hill, New York, 1979.
4. ••

PROBLEM

9-1. Consider the process shown in Fig. P9-1. In the reactor the principal reaction is $A + 2B \rightarrow P$; two other reactions, $A + 2B \rightarrow$ inert and $A \rightarrow$ heavies, also occur but at a lesser rate. All the reactions occur in the gas phase. Enough cooling is accomplished in the cooler to condense and separate the heavies. The gases are separated in the separation column. The gases leaving the column contain A, B, and inerts. The purge is manipulated to maintain the composition of inerts in the recycle stream at some desired value, 1 mol %. In the recycle line there is a temperature transmitter, TT1; a volumetric flow transmitter, FT3; and two continuous infrared analyzers. One of the analyzers, AT1, gives the mole fraction of A, y_{AR}, and the other analyzer, AT2, gives the mole fraction of B, y_{BR}. The process has been designed to minimize the pressure drop between the column and the compressor. The reactants A and B are pure components and are assumed to be delivered to the valves at constant pressure and temperature.

(a) Design a control scheme to control the composition of inerts in the recycle stream at 1 mol %.

(b) Design a control scheme to control the supply pressure to the compressor. It is also very important to maintain the molal ratio of B to A entering the compressor at 2.6. There is one infrared analyzer, AT4, at the exit of the compressor that provides a signal indicating this ratio.

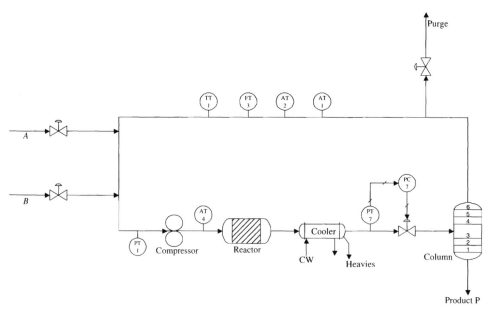

Figure P9-1 Process for Problem 9-1.

APPENDIX A

CASE STUDIES

In this appendix we present a series of design case studies that provide the reader with an opportunity to design process control schemes. The first step in designing control systems for process plants is deciding which process variables must be controlled. This decision should be made by the process engineer who designed the process, the instrument or control engineer who will design the control system and specify the instrumentation, safety engineers, and the operating personnel who will run the process. This is certainly very challenging and requires team effort. The second step is the actual design of the control system. In the case studies that follow, the first step has been done. It is the second step that is the subject of these case studies. Please note that like any design problem, these problems are open-ended. That is, there are multiple answers.

Case 1: Ammonium Nitrate Prilling Plant Control System [1]

Ammonium nitrate is a major fertilizer. The flowsheet shown in Fig. A-1 shows the process for its manufacture. A weak solution of ammonium nitrate (NH_4NO_3) is pumped from a feed tank to an evaporator. At the top of the evaporator there is a steam ejector vacuum system. The air fed to the ejector controls the vacuum drawn. The concentrated solution is pumped to a surge tank and then fed into the top of a prilling tower. The development of this tower is one of the major postwar developments in the fertilizer industry. In this tower the concentrated solution of NH_4NO_3 is dropped from the top against a strong updraft of air. The air is supplied by a blower at the bottom of the tower. The air chills the droplets in spherical form and removes part of the moisture, leaving damp pellets or prills. The pellets are then conveyed to a rotary dryer, where they are dried. They are then cooled, conveyed to a mixer for the addition of an antisticking agent (clay or diatomaceous earth), and bagged for shipping.

199

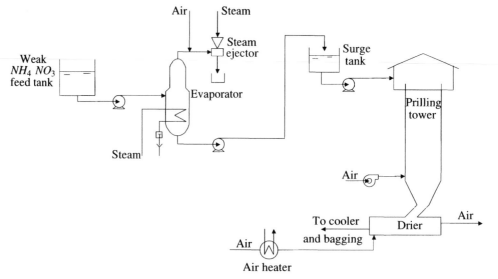

Figure A-1   NH$_4$NO$_3$ process.

How would you control the production rate of this unit? Design the system to implement the following:

· Control the level in the evaporator.
· Control the pressure in the evaporator. This can be accomplished by manipu-lating the flow of air to the exit pipe of the evaporator.
· Control the level in the surge tank.
· Control the temperature of the dried pellets leaving the dryer.
· Control the density of the strong solution leaving the evaporator.

Be sure to specify the action of valves and controllers.

If the flow to the prilling tower varies often, it may also be desired to vary the air flow through the tower. How would you implement this?

Case 2:   Natural Gas Dehydration Control System

The process shown in Fig. A-2 is used to dehydrate the natural gas entering the absorber using a liquid dehydrant (glycol). The glycol enters the top of the absorber and flows down the tower countercurrent to the gas, picking up the moisture in the gas. From the absorber, the glycol flows through a cross-heat exchanger into the stripper. In the reboiler, at the base of the stripper, the glycol is stripped of its mois-ture, which is boiled off as steam. This steam leaves the top of the stripper and is condensed and used for the water reflux. This water reflux is used to condense the glycol vapors that might otherwise be exhausted along with the steam.

The process engineer who designed the process has decided that the following must be controlled.

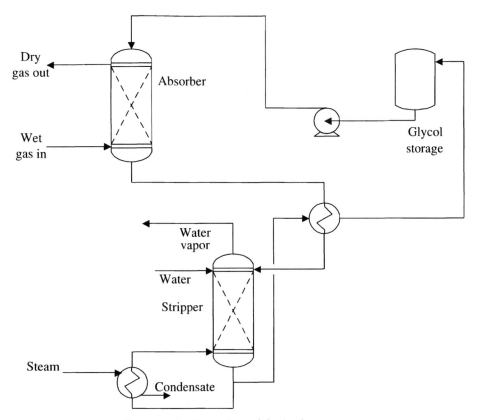

Figure A-2   Natural gas dehydration system.

- The liquid level at the bottom of the absorber
- The water reflux into the stripper
- The pressure in the stripper
- The temperature in the top third of the stripper
- The liquid level at the bottom of the stripper
- Efficient absorber operation at various throughputs

Design the control system to accomplish the desired control.

Case 3:   Sodium Hypochlorite Bleach Preparation Control System [1]

Sodium hypochlorite (NaOCl) is formed by the reaction

$$2NaOH + Cl_2 \rightarrow NaOCl + H_2O + NaCl$$

The flowsheet in Fig. A-3 shows the process for its manufacture.

Dilute caustic (NaOH) is prepared continuously to a set concentration (15% solution) by water dilution of a 50% caustic solution and stored in an intermediate

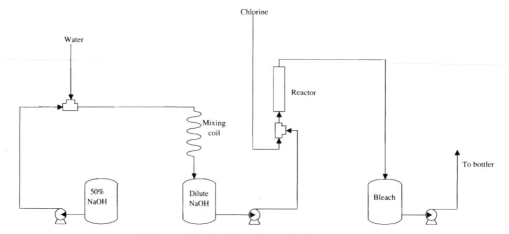

Figure A-3   Sodium hypochlorite process.

tank. From this tank the solution is then pumped to the hypochlorite reactor. Chlorine gas is introduced into the reactor for the reaction.

How would you set the production rate from this unit? Design the control system to accomplish the following:

· Control the level in the dilution tank.
· Control the dilution of the 50% caustic solution. The concentration of this stream is to be measured by a conductivity cell. When the dilution of this stream decreases, the output from this cell increases.
· Control the level in the bleach liquor storage tank.
· Control the ratio of excess NaOH/available Cl_2 in the outlet stream from the hypochlorite reactor. This ratio is measured by an ORP (oxidation–reduction potential) technique. As the ratio increases, the ORP signal also increases.

Specify the action of valves and controllers.

For safety reasons, when the flow of caustic solution from the dilute caustic tank to the reactor fails, the flow of chlorine must be stopped immediately. Design and explain this scheme.

Case 4:   Control Systems in the Sugar Refining Process

The process units shown in Fig. A-4 form part of a process to refine sugar. Raw sugar is fed to the process through a screw conveyor. Water is sprayed over it to form a sugar syrup. The syrup is heated in the dilution tank. From the dilution tank the syrup flows to the preparation tank where more heating and mixing are accomplished. From the preparation tank the syrup flows to the blending tank. Phosphoric acid is added to the syrup as it flows to the blending tank. In the blending tank, lime is added. This treatment with acid, lime, and heat serves two purposes. The first is that of clarification; that is, the treatment causes coagulation and precipitation of

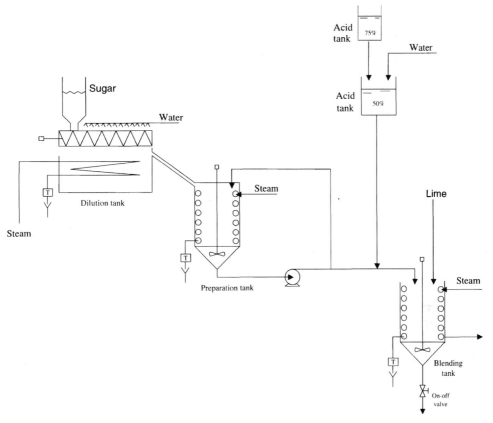

Figure A-4   Sugar refining process.

the no-sugar organics. The second purpose is to eliminate the coloration of the raw sugar. From the blending tank the syrup continues to the process.

How would you control production rate? The following variables are thought to be important to control.

- Temperature in the dilution tank
- Temperature in the preparation tank
- Density of the syrup leaving the preparation tank
- Level in the preparation tank
- Level in the 50% acid tank (the level in the 75% acid tank can be assumed constant)
- Strength of the 50% acid (the strength of the 75% acid can be assumed constant)
- Flow of syrup and 50% acid to the blending tank
- pH of the solution in the blending tank
- Temperature in the blending tank

The blending tank requires only a high-level alarm. The flowmeters used in this process are magnetic flowmeters. The density unit used in the sugar industry is °Brix, which is roughly equivalent to the percentage of sugar solids in the solution by weight. Design the control systems necessary to control all of the variables above. Show the action of control valves and controllers.

Case 5: Sulfuric Acid Process

Figure A-5 shows a simplified flow diagram for the manufacture of sulfuric acid (H_2SO_4). Sulfur is loaded into a melting tank, where it is kept in the liquid state. From this tank the sulfur goes to a burner, where it reacts with the oxygen in the air to produce SO_2 by the reaction

$$S_{(l)} + O_{2(g)} \rightarrow SO_{2(g)}$$

From the burner the gases pass through a waste heat boiler where the heat of reaction of the reaction above is recovered by producing steam. From the boiler the gases then pass through a four-stage catalytic converter (reactor). In this converter the following reaction takes place:

$$SO_{2(g)} + \frac{1}{2}O_{2(g)} \rightarrow SO_{3(g)}$$

From the converter, the gases go to an absorber column where the SO_3 gases are absorbed by dilute H_2SO_4 (93%). The water in the dilute H_2SO_4 reacts with the SO_3 gas, producing H_2SO_4:

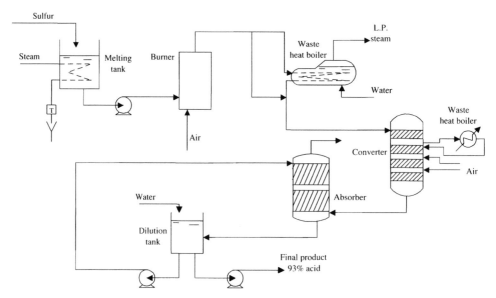

Figure A-5 Sulfuric acid process.

$$H_2O_{(l)} + SO_{3(g)} \rightarrow H_2SO_{4(l)}$$

The liquid leaving the absorber, concentrated H_2SO_4 (98%), goes to a circulation tank where it is diluted back to 93% using H_2O. Part of the liquid from this tank is then used as the absorbing medium in the absorber.

How would you set the production rate for this plant? The following variables are thought to be important to control:

- Level in the melting tank
- Temperature of sulfur in the melting tank
- Air to the burner
- Level of water in the waste-heat boiler
- Concentration of SO_3 in the gas leaving the absorber
- Concentration of H_2SO_4 in the dilution tank
- Level in the dilution tank
- Temperature of the gases entering the first stage of the converter

Design the necessary control systems to accomplish the above. Be sure to specify the action of valves and controllers.

Case 6: Fatty Acid Process

Consider the process shown in Fig. A-6. The process hydrolyzes crude fats into crude fatty acids (CFAs) and dilute glycerine using a continuous high-pressure fat split-

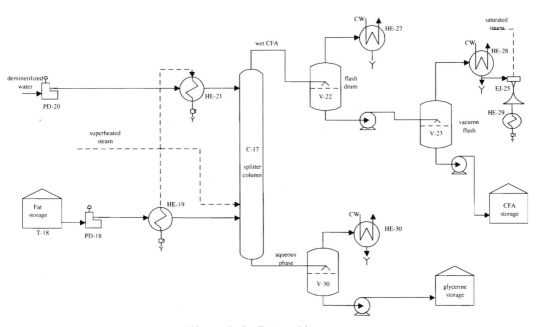

Figure A-6 Fatty acid process.

ter column (C17). The main product is high-quality CFAs. The CFA quality is primarily a function of the acid value. In the column the following reaction takes place:

$$
\begin{array}{c}
\mathrm{CH_2OCOR} \\
| \\
\mathrm{CHOCOR'} \\
| \\
\mathrm{CH_2OCOR''}
\end{array}
+ 3\mathrm{H_2O}
\xrightarrow[\text{T and P}]{\text{high}}
\begin{array}{c}
\mathrm{CH_2OH} \\
| \\
\mathrm{CHOH} \\
| \\
\mathrm{CH_2OH}
\end{array}
+
\begin{array}{c}
\mathrm{RCH{=}O{-}OH} \\
\mathrm{R'CH{=}O{-}OH} \\
\mathrm{R''CH{=}O{-}OH}
\end{array}
$$

Triglyceride   Water   Glycerine   Mixed Fatty Acids

Fat is stored (T18) at 120°F and pumped into the column using a positive displacement pump (PD18). The fat is preheated (HE19) to 400°F with superheated steam before it enters the column. The column operates continuously at 700 psig and 500°F with a crude fat feed rate of 25,000 lb/hr.

Demineralized water is pumped into the column using a positive displacement pump (PD20). The water is preheated (HE21) to 500°F. Excess water is required to assure complete hydrolysis of the crude fat.

Superheated steam at 800 psig and 700°F is sparged directly into the column. The steam provides heat and mixing to break up the fat.

The splitter is basically a countercurrent contactor. Water feed at the top has a higher specific gravity than the CFAs. Crude fat feed at the bottom is insoluble in water and rises as the water migrates down the column. The glycerine produced by the reaction is soluble in water and increases the specific gravity of the aqueous phase.

An interface forms in the column. Above the interface the material is mostly fat and CFAs. Below the interface is mostly an aqueous phase of water and glycerine. The best operation of the column is achieved when this interface is located near the steam sparger. If the interface level is low, the amount of CFA in the aqueous phase increases. If the level is too high, fat dispersion into the water is lost and incomplete hydrolysis results. High temperature is required to produce the hydrolysis reaction, but boiling must be avoided, as this condition causes the aqueous phase to rise and upset the column.

The material removed overhead contains CFAs and a small amount of water. This wet CFA is a light brown milky material. The overhead product is dried by a two-step flash process. The sensible heat of the material is enough to dry the material without heat. The material is sprayed into the first vessel (V22), and most of the water evaporates. The overhead water is condensed (HE27). The resulting CFA is then sent to a vacuum flash (V23) to dry the material fully. A steam jet ejector (EJ25) is used to draw vacuum. The overhead water in the vacuum flash is condensed in the precondenser (HE28) and sent to the sewer. The noncondensables from the precondenser are pulled through the steam jet ejector and the motive steam condensed in the barometric condenser (HE29). The vacuum flash tank should be operated at 100 mmHg. The ejector is significantly oversized for normal duty and consumes 2500 lb/hr of 150-psig saturated steam. Very low pressure will cause low-molecular-weight elements of the CFAs to vaporize and foul the precondenser. Loss of vacuum will allow wet CFAs to be stored in the tank, which will cause problems in downstream processes.

The aqueous phase is removed from the bottom of the column and should be 20 weight percent glycerine dissolved in water. Like the CFAs, the aqueous phase is flashed at atmospheric pressure (V30). Any fatty material in the aqueous phase makes purification of the glycerine very difficult. Excess water in the aqueous phase requires additional energy in the glycerine purification. Glycerine is a clear colorless liquid.

Prepare a detailed instrument diagram to control:

· Production rate from the process
· Level in the splitter column
· Level in all flash tanks
· Pressure in the column
· Pressure in the vacumn flash
· Temperature in the splitter column
· Temperature in the heaters

All instruments shown should be tagged and the normal operating value and proposed range of the instrument provided.

REFERENCE

1. Foxboro Co. *Application Engineering Data AED 288-3*, January 1972.

APPENDIX B

PROCESSES FOR DESIGN PRACTICE

In this appendix we describe three processes presented in the CD to practice the material presented in the book. Specifically, these processes may be used to practice tuning feedback controllers; process 1 may also be used to design a feedforward controller using the block diagram method; process 2 may also be used to tune a two-level cascade system. The program used to develop the processes is Labview, a product of National Instruments.

Installing the Programs

Insert the CD into the drive, address the drive, the CD may start running by itself, if not click on Setup, and follow instructions. Once everything is completed the programs will reside in C:\Program Files\Control. Figure B-1 shows the menu that will appear once you run Control. Once you open a process it will be running. You should wait just a few minutes for the process to reach steady state before you start working.

Process 1:   NH₃ Scrubber

The process shown in Fig. B-2 is used to practice tuning a feedback controller and to design a feedforward controller. The process is a scrubber in which HCl is being scrubbed out of air by a NaOH solution. The analyzer transmitter in the gas stream leaving the scrubber has a range of 25 to 150 ppm. The HCl–air mixture is fed to the scrubber by three fans. Fan 1 feeds 50 cfm, and fans 2 and 3 feed 25 cfm each. These fans are turned on/off simply by clicking on them with the mouse. The set point is changed by either double clicking on the number and entering the new set point, or by clicking on the up/down arrows next to the numerical value; each click changes the set point by 1 ppm. The controller's action is set by clicking on the switch, indicating reverse (REV) or direct (DIR). When in the manual mode, the

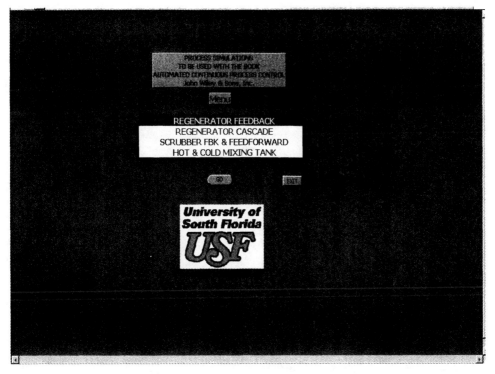

Figure B-1   Menu.

controller's output is set by either double clicking on the number and entering the output, or by clicking the up/down arrows next to the value; each click changes the output by 1%. The user also has the capability of selecting the PV tracking option. The controller's tuning terms—gain, reset, and rate—are set by either by double clicking on the number itself and entering the value, or by clicking the up/down arrows. This is also the way to set the terms of the feedforward controller—gain, lead time constant, lag time constant, and dead time.

The top chart shows the set point (in red) and the outlet ppm (in blue) of the gas stream. The bottom chart shows the controller's output (in red) and the feed flow to the scrubber (in blue). To re-range the charts double click the numerical value in the axis and typing the new value.

Tuning the Feedback Controller. To tune the feedback controller, that is, to find K_C, τ_I, and τ_D, we must first find the process characteristics, process gain K, time constant τ, and dead time t_0. In Chapter 2 we explain that to obtain these characteristics, a process reaction curve is necessary. As explained in that chapter, to obtain this curve we first introduce a step change in the controller's output and record the ppm of the NH_3 leaving the scrubber. Unfortunately, there is no recorder to record the ppm. Thus you will have to generate a table of the NH_3 ppm versus and graph these data. We recommend that the ppm be read every 5 sec. You should read the ppm until a steady state is achieved again. To generate the step change in the controller's output, double-click on the number, type the new desired output, and press

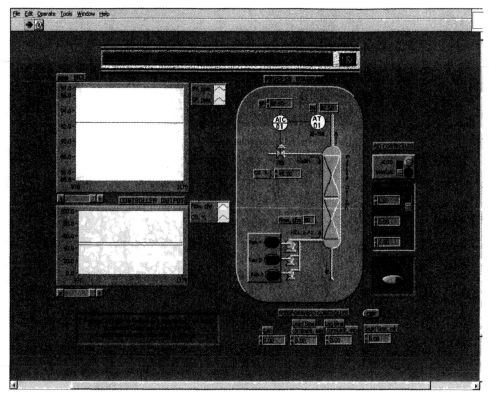

Figure B-2 Scrubber.

"enter". You may read Section 2-5 again to review the questions related to this process testing.

Once the process characteristics are obtained and the controller tuned using the formulas of Chapter 3, the tuning can be tested by changing the set point or by introducing a disturbance by starting or stopping a fan.

Designing the Feedforward Controller. It is known that for this scrubber unit the input flow can vary much and very often. That is, it is common to have one fan working (e.g., fan 2 with a flow of 25 cfm, and suddenly the other two fans turn on, to have a total input flow of 100 cfm. When this occurs, the outlet HCl will increase until the feedback controller can take hold by adding more NaOH to bring the HCl back to set point. We must be sure that the HCl ppms do not violate the EPA regulation. Thus, under normal operation the set point to the controller must be set low enough to ensure staying within regulation under this upset condition. Although the low set point guarantees not violating the regulation during any upset condition, it costs extra money during normal operation because it requires extra NaOH flow. If we could provide a tighter control, less deviation from set point under upset conditions, we could raise the set point and thus save a portion of the extra NaOH.

Feedforward control can provide this tighter control. The idea is to measure the flow entering the scrubber, and if this flow changes, manipulate the NaOH valve. That is, do not wait for a deviation in HCl before taking action. There is a flow sensor/transmitter with a range of 0 to 100 cfm measuring the input flow to the scrubber.

Chapter 6 shows that a way to design a feedforward controller is to use a first-order-plus-dead time transfer function (K, τ, and t_0) describing how the disturbance (feed to the scrubber in this case) affects the controlled variable (outlet HCl ppm in this case), and another first-order-plus-dead time transfer function describing how the manipulated variable affects the controlled variable. The former transfer function is obtained by clicking on/off one of the fans to generate a change in input flow, recording (every 5 sec) the HCl ppm value, graphing the response curve, and using the two-point method described in Chapter 2. The latter transfer function was obtained to tune the feedback controller. Once both transfer functions are obtained, Eq. (7-2.5) is used to tune the feedforward controller.

Once the feedforward controller is designed, it can be tested. We recommend the testing to proceed the following way. Under feedback control, one of the fans should be turned on or off and this control performance be used as a baseline to compare the feedforward performance. Then, under steady-state feedforward/feedback control, generate the same disturbance and compare the control obtained with that of feedback. Repeat, but this time add the lead/lag unit. Finally, do the same, adding the dead-time compensator if needed.

Process 2:   Catalyst Regenerator

A catalyst is used in a hydrocarbon reaction. As the reaction proceeds, some carbon deposits over the catalyst, poisoning the catalyst. After enough carbon has deposited, it poisons the catalyst completely. At this moment it is necessary to stop the reaction and regenerate the catalyst. This regeneration consists in blowing hot air over the catalyst so that the oxygen in the air reacts with the carbon to form carbon dioxide, and in so doing to burn the carbon. Figure B-3 shows the regeneration process. Ambient air is first heated in a small furnace and then flows to the regenerator, which is full of catalyst. Manipulating the fuel flow controls the temperature in the catalyst bed. The set point is changed by either double clicking on the number and entering the new set point, or by clicking on the up/down arrows next to the numerical value; each click changes the set point by 1°F. The controller's action is set by clicking on the switch indicating reverse (REV) or direct (DIR). The controller's output is set, when in the manual mode, by either double clicking on the number and entering the output, or by clicking the up/down arrows next to the value; each click changes the output by 1%. The controller's tuning terms—gain, reset, and rate—are set by either by double clicking on the number itself and entering the value, or by clicking the up/down arrows.

Tuning the Feedback Controller. To tune the feedback controller we must first find the process characteristics. Section 2-5 we explain that to obtain the process characteristics, a process reaction curve is necessary. To obtain this curve, we introduce a step change in the controller's output and record the temperature in the regenerator. Unfortunately, there is no recorder to record this temperature. Thus you will have to generate a table of the temperature in the regenerator versus time

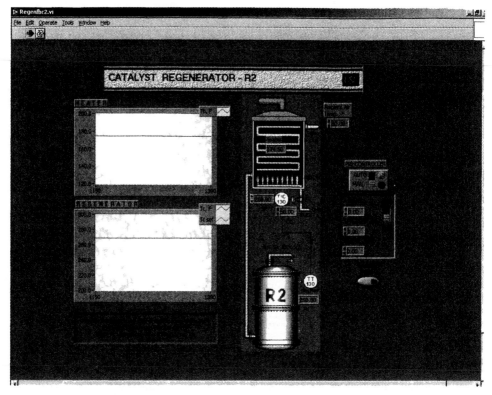

Figure B-3 Catalyst regenerator—feedback control.

and graph these data. We recommend that the temperature be read every 5 sec. You should read the temperature until a steady state is achieved again. To generate the step change in the controller's output, double-click on the number, type the new desired output, and press "enter." You may reread Section 2-5 to review the questions related to this process testing.

Once the process characteristics are obtained and the controller tuned using the formulas of Chapter 3, the tuning can be tested by changing the set point or introducing a disturbance. The temperature of the air entering the furnace can be changed to induce a disturbance. We recommend changing this inlet temperature, by 10°F and recording the largest deviation from the set point. Next you implement a cascade control scheme, and can then compare the control performance given by feedback and cascade.

Tuning Cascade Controllers. In Chapter 4 we explained how to tune cascade controllers. Figure B-4 shows the cascade control implemented in the regeneration process. The first step is to obtain the process reaction curve for the secondary variable and the process curve for the primary variable. Both curves are generated by the same step change in the secondary controller's output; this is the controller connected to the final control variable. This time you will have to generate two tables, the temperature in the regenerator (primary controlled variable) versus time, and the temperature leaving the furnace (secondary controlled variable) versus time.

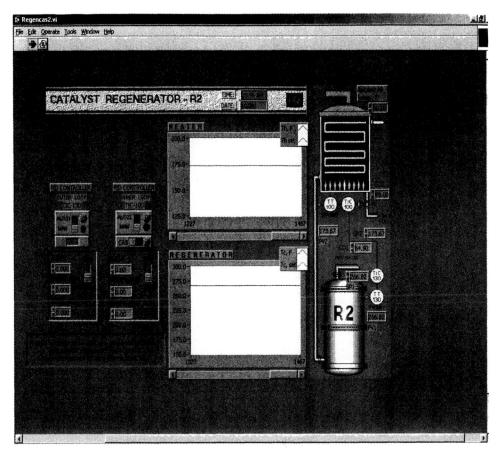

Figure B-4   Catalyst regenerator—cascade control.

Remember, in taking the data you should read the variables every 5 sec after changing the controller's output. From the temperature in the regenerator versus time graph, you obtain K_1, τ_1, and t_{01}, and from the temperature leaving the furnace versus time graph, you obtain K_2, τ_2, and t_{02}. Using the last set of terms the secondary controller is tuned as a simple feedback controller (tuning formulas presented in Chapter 3). Using all the terms and the tuning of the secondary, the primary controller is tuned using either Table 4-2.1 or 4-2.2. Once you have tuned both controllers, you should try the secondary first to make sure that it works fine by itself. Once this is done, you can set the secondary controller in cascade (remote set point) and the primary controller in automatic. A good test to perform is to change the inlet air temperature to the furnace and record the largest deviation from set point. You can then compare this deviation with the one you obtained under simple feedback control.

Process 3:   Mixing Process

Figure B-5 shows the schematic of a tank where cold water is mixed with hot water. The valve in the cold water pipe is manipulated to control the temperature of the

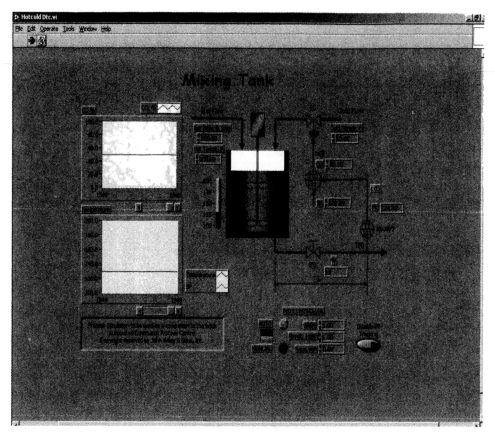

Figure B-5 Mixing tank.

water leaving the exit pipe. You may assume that a level controller (not shown) maintains perfect level control. The hot water flow and temperature and the cold water temperature act as disturbances to the process. The set point is changed by either double clicking on the number and entering the new set point, or by clicking on the up/down arrows next to the numerical value; each click changes the set point by 1°F. The controller's output is set, when in the manual mode, by either double clicking on the number and entering the output, or by clicking the up/down arrows next to the value; each click changes the output by 1%. The user also has the capability of selecting the PV tracking option. The controller's tuning terms—gain, reset, and rate—are set by either by double clicking on the number itself and entering the value, or by clicking the up/down arrows.

Tuning the Feedback Controller. Tune the feedback controller by the method presented in Chapters 2 and 3, and used in the other processes. Note the units of the reset time. A very interesting disturbance is the flow of hot water. Once the temperature controller is tuned, start to decrease the hot flow by $25\,\mathrm{lb_m/min}$ at a time and watch the control performance.

INDEX

action, 3
analog, 5

block diagrams, 127
boiler control, 83
boiler level control
   single-element, 168
   three-element, 169
   two-element, 168

cascade control
   primary variable, 63
   secondary variable, 64
cascade controllers, tuning, 65
combustion control, 83
computing algorithms, 74
control
   automatic, 2, 4
   cascade, 61
   closed-loop, 4
   constraint, 88
   cross-limiting, 86
   feedback, 2, 6
   feedforward, 8, 142
   manual, 4
   override, 88
   ratio, 80
   regulatory, 4
   selective, 92
   servo, 5
controller, 2, 38
   action, 38
   derivative time, 48

gain, 40
master, 63
offset, 41
primary, 63
proportional (P), 40
proportional band, 43
proportional-derivative (PD), 50
proportional-integral (PI), 44
proportional-integral-derivative (PID), 48
rate, 48
reset, 44
slave, 64
stability, 132
tuning, 53
cross-limiting control, 86

dead time, 21
dead time compensation, 151, 174
   Dahlin's controller, 176
   Smith Predictor, 174
decision, 3
digital, 5
distributed control systems (DCS's), 38, 74
disturbance, 4

external reset feedback, 90

feedforward control, 8, 142
   dynamic, 153
   steady-state, 153
final control element, 2
forcing function, 14

215

gain, 17

lead/lag, 151, 155
level
  average control, 59
  tight control, 58

master controller, 63
multivariable process control, 180
  decoupling, 194
  pairing, 181
  stability, 191
  tuning, 192
measurement, 3

noise, 33

offset, 41

process, 11
  characteristics, 11
  dead time, 21
  first-order, 24
  first-order-plus-dead-time (FOPDT), 24
  gain, 17
  higher-order, 26
  integrating, 14
  multicapacitance, 24
  nonlinearities, 23
  nonself-regulating, 13
  open loop unstable, 14
  reaction curve, 29
  self-regulating, 13
  single capacitance, 14
  time constant, 20

programming
  block oriented, 76
  software oriented, 76

relative gain analysis, 184
reset feedback, 90
reset windup, 50
responding variable, 14

sensor, 2
set point, 4
  local, 64
  remote, 64
signals, 5
slave controller, 64
stability, 132

time constant, 20
tracking, 73
transducer, 5
transfer function, 24
transmitter, 2, 28
tuning
  cascade controller, 65
  controller synthesis method, 55
  flow loops, 56
  lambda method, 55
  level loops, 57
  Ziegler-Nichols, 53, 55
two-point method, 29

ultimate gain, 54, 136
ultimate period, 54, 136
upset, 4

variable
  controlled, 3
  manipulated, 4

Printed in the United States of America/BNB